Packet Radio: What? Why? How?

Articles and Information on General Packet Radio Topics

Edited by

Greg Jones, WD5IVD

Published by: Tucson Amateur Packet Radio Corporation

Printed in the USA

ISBN: 0-9644707-0-5

$ 9.00 in USA

First Edition
First Printing, 1995

Trademark Notices

Apple is a trademark of Apple Computer Corporation

AMSAT is a trademark of The Radio Amateur Satellite Corporation

Commodore 64 is a trademark of the Commodore Business Machines, Inc

IBM is a trademark of International Business Machines Corporation

KAM, KA-Node, KPC-1, KPC-2, KPC-3, are trademarks of Kantronics

Macintosh is a trademark of Apple Computer Corporation

MS-DOS is a trademark of Microsoft Corporation

PAKRATT, PKT-1, PK-80, PK-64, PK-87, PK-232, DSP-2232 are trademarks of Advanced Electronic Applications, Inc

Teletype is a trademark of the Teletype Corporation

UNIX is a trademark of Bell Labs

Introduction

Welcome to the wonderful world of packet radio. The TAPR Packet Radio: What? Why? How? book provides an introduction to and general information on packet radio. The concept of the book is to provide information to beginning and intermediate packet operators. This is the first version of this book, so please send in your comments on how we can improve it in the future.

I would like to thank all the people who made contributions and had input on this version of the book. We hope everyone enjoys this reference.

73 / Greg Jones - WD5IVD

Send comments, suggestions, or corrections:

TAPR
ATTN : "Packet Radio: What? Why? How?"
8987-309 E Tanque Verde Rd #337
Tucson, Az 85732

Tucson Amateur Packet Radio
8987-309 E. Tanque Verde Rd #337
Tucson, Arizona • 85749-9399
Office: (817) 383-0000 • Fax: (817) 566-2544
Non-Profit Research and Development Corporation

Contents

TAPR on the Internet

TAPR can be reached via the Internet. The Automated Information Server that TAPR provides allows anyone to request information on TAPR, products, newsletters, and lots of other files. To find out more about this service, send an e-mail message to listserv@tapr.org with the subject line "Request" and the following text in the body of the message:

help	(for a brief set of instructions)
index -all	(for a list of all file by topic area)
get tapr taprinfo.txt	(or info on TAPR)

In the above message example, "help" retrieves a brief set of instructions for info, "index -all" retrieves a list of available files by topic and "taprinfo.txt" retrieves a text file containing information on TAPR and what TAPR is all about. If you want to retrieve several text files with one message, use a separate line for each "get filename" request.

TAPR Special Interest Groups

TAPR maintains several SIGs in various areas. Send an e-mail message to listserv@tapr.org with the subject line "Request" and the following text in the body of the message:

lists	(for list of mail groups on TAPR.ORG)
information listname	(where listname is the name of the mail group)

When you are ready to subscribe, send e-mail to listserv@tapr.org with the following command in the message text:

subscribe <list> <your name>

FTP

The TAPR Software Library is available via anonymous FTP. You can access the library by ftp to 'ftp.tapr.org'. Login in as 'anonymous', with a password of 'your_account@internet_address'. In addition, 'ftp.hereford.ampr.org' also has the library available.

World-Wide-Web!

The series is accesible through the Packet Radio Home Page at this URL:

http://bb.iu.net/infomotion/taprhome.html
http://bb.iu.net/infomotion/pkthome.html

Why Packet Radio?

Could be, You are Missing Something Fun.

Greg Jones, WD5IVD

Packet radio has been around since the mid-1960's, but was first seen on the amateur bands in 1978 through research done by the Vancouver Amateur Digital Communication Group (VADCG). This was then followed by TAPR (Tucson Amateur Packet Radio) with the creation of the TNC-1 in 1982 and then the TNC-2 in '84-'85. Ten years ago, the packet radio revolution ignited when TAPR sold over a thousand TNC-2 kits. The TNC-2 was what was needed to make this mode, that a few experimenters were playing with, into something that every amateur could enjoy. From its humble beginnings, where it was good luck to have more than three packet operators in the same city, packet radio now has thousands of amateurs using it daily, various manufacturers making and selling TNCs (terminal Node Controllers), and over a hundred thousand TNCs having been sold to date. What growth! No other mode of amateur radio has seen such explosive growth!

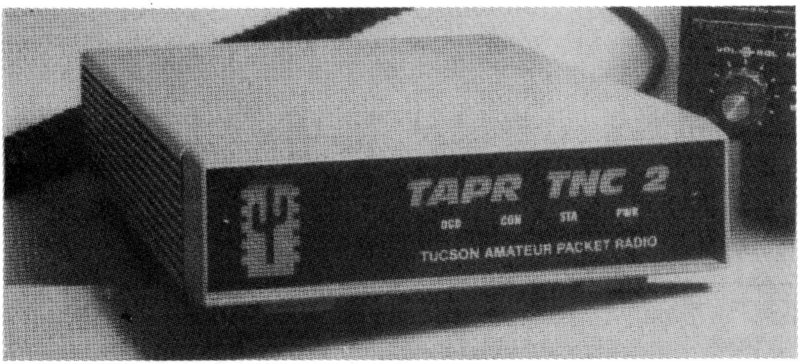

Like the title says, 'Why Packet Radio ?' Like any mode in the amateur service, it provides a group of amateurs with a way of having fun and meeting one of our primary aims—'improving the radio art.' Packet radio was a new mode in the early 80's that many of the outstanding amateur experimenters worked on and developed. The result, ten years later, is something that provides a lot of different operating opportunities. No longer is it just packet radio, but now it is bulletin board systems, DX Clusters, chat bridges, networking, emergency communications, satellite operations and much more. But what are these ? and is one of these, something that you want to do? How do you know? Let's start off with a basic question.

• What is packet radio?

The good thing about packet radio is that you don't really have to know a lot about how it works or find it necessary to memorize a whole new set of technical terms. Find a friend who is using packet, buy your TNC (terminal node controller), hook up your unit, and then ask for help. The nice thing these days is that almost every town has someone on packet radio who can help. A basic TNC allows your computer to use your radio to talk to another computer, thus combining two popular hobbies—computers and radios. The cost of the TNC is going to depend on what you want to do. The question that you should ask before 'What TNC do I want?' is 'Why do I want to invest in new equipment?' Let's spend the rest of the article talking a little about the most popular uses of packet radio. After you read this, find someone locally on packet and ask for a demonstration. Since amateur packet radio is different in every fifty mile radius, then what I can do here in Austin, Texas is going to be different from what you can do where you are. Find out what you are going to do before spending your money, unless you want to blaze a new trail of services in your area.

Packet Bulletin Board Systems (BBS): Most cities have one or more packet Bulletin Board Systems, or BBS for short. BBSs do two main things: send and receive personal messages for their local users (like yourself) and send and receive messages or bulletins intended for people locally or around the world. Since the BBS is part of a national system of other BBSs, it has the ability to pass information or messages to any other BBS in the US or the world.

Bulletins listed from W5SYT BBS in Austin Nov 11, 1994

MSG #	TR	SIZE	TO	FROM	@BBS	DATE	TITLE
18901	B#	1268	SALE	KC6WIM	ALLUS	941111	ASTRON POWER SUPPLY
18900	B#	1269	ALL	N6ZUI	ALLUS	941111	SALE ICOM R100
18897	B#	1299	ALL	N6ZUI	ALLUS	941111	SALE ICOM R-71A
18896	B#	1390	SALE	N6MEL	ALLUS	941111	LAPTOP FOR PACKET
18895	B#	1399	SALE	KC6WIM	ALLUS	941111	YAESU FT-33R 220MHZ HT
18894	B#	1435	ALL	W6IOW	ALLUS	941111	ATLAS 400X
18893	B#	1579	GTOR	WB6AAM	ALLUS	941111	GTOR on 160 meters??!!
18892	B#	1284	ALL	KC5BFA	ALLUS	941111	QRP RIG NEEDED!!
18891	B#	1097	SALE	KE4MG	ALLUS	941111	RE ht's for sale
18890	B#	1298	SALE	WA4WEB	ALLUS	941111	Sale:KX-P1180 PRINTER
18888	B#	1221	SALE	WA4WEB	ALLUS	941111	SALE:AEA PK-232
18887	B#	1204	WANT	KD4AJ	ALLUS	941111	Kenwood TS-850S
18886	B#	940	DOOM	W3BRZ	ALLUS	941111	DOOM 2 HELP??
18885	B#	1753	SALE	N8XKX	ALLUS	941111	MC-50 FOR SALE
18884	B#	1434	SALE	KE4MG	ALLUS	941111	IBM PC-jr and acc

Enter command: A,B,C,D,G,H,I,J,K,L,M,N,P,R,S,T,U,V,W,X,Y,?,* >

This allows you to send messages to friends locally, to someone located in the next state, or to someone on the other side of the world. The second thing that BBSs do is pass local and national bulletins, which are messages intended to be read by everyone. In this way, amateurs can read the latest messages about the ARRL, AMSAT, TAPR, propagation, DX, and other bulletins on varied topics. Message passing is the primary purpose of a BBS system, but BBSs can also support callbook programs, help references, Internet access, and more. Operators of BBS systems are a good place to start when you first get on the air. Because of the service they provide, they have to know how packet is working in the local area.

Keyboard-to-Keyboard: Like other amateur modes (SSB, FM, etc), packet radio can be used to talk to other amateurs directly. Amateurs can talk to each other simultaneously using their keyboards when they can directly communicate with each other. With the use of networks (see a little later), amateurs can talk at a distance beyond the reach of their own stations by using the network. Keyboard-to-keyboard communications is one of the least frequent methods of packet communications, because amateurs are rarely on packet at the same time. Many packet operators send electronic mail using either personal mailboxes or a local BBS. In this way, messages are read when the amateur is on the air. Another limitation to direct keyboard-to-keyboard packet is that you can only talk to one packet station at a time — no easy way to hold round-table discussions like on a voice repeater. Some areas support chat or conference bridges, which allow for more than one amateur to talk to each other — much like a voice repeater. If a chat is supported over a network, then you can talk to someone as far away as the network reaches.

DX Packet Cluster: Many cities have DX (foreign amateur) spotting nodes or networks. HF (High-Frequency) operators connect to their local DX Packet Cluster in order to receive reports on the latest DX. This type of packet came about from those interested in 'chasing' DX. Many amateurs like to frequent the HF bands looking for rare international operators to contact. A DX Cluster allows many HF operators to be connected over packet radio at the same time while operating HF and hunting for DX. When someone finds a DX station, they send a packet message to the DX Cluster, which then sends the information to all other

Spots from the DX Cluster Network in Texas on March 7th, 1994

14022.4	UQ2GP	7-Mar-1994	0421Z		<NJ5N>
14184.8	AP2SQ	7-Mar-1994	0336Z	RX 205-215	<W5PLN>
28023.4	YB0BAQ	7-Mar-1994	0259Z	OP TOM	<NJ5N>
14030.5	U1PR	7-Mar-1994	0248Z		<N5GM>
14028.5	UL7XE	7-Mar-1994	0243Z	OP JURI	<NJ5N>
3792	KP2A/KP5	7-Mar-1994	0212Z		<N5HRG>
7005.1	K4LTA/J8	7-Mar-1994	0210Z	up 1	<KF5PE>
21006.5	PZ1DY	7-Mar-1994	0053Z		<N5RZ>
21011	TU4CO	6-Mar-1994	2238Z		<WS5E>
14009	TR2GA	6-Mar-1994	2228Z		<WS5E>
28545.0	SP9AI	6-Mar-1994	1539Z		<N5AU>
28540.0	HA0KLE	6-Mar-1994	1537Z		<N5AU>
28477.9	UB4QYA	6-Mar-1994	1517Z		<N5AU>
28529.2	UV3QTR	6-Mar-1994	1509Z		<N5AU>
14207.7	VU2XX	6-Mar-1994	1358Z		<KA5BJY>

packet operators using the DX Cluster. In this way, you have several stations monitoring the band, looking for DX. Often an amateur will 'spot' (hear) a DX station and then distribute the DX report almost instantly. DX Clusters allow everyone to operate many more hard-to-find DX stations in one evening than was possible operating by oneself. Some amateurs have been known to attain enough contacts to qualify for DXCC in a matter of weeks. One point though, if your HF station is not a 'big-gun', then it is sometimes best to operate the DX station before posting your spot for others to find. There is a good chance that a pile-up will occur as soon as you make your spot to the DX Cluster and then you will not be able to work the DX station that you found!

RACES/ARES/NTS and Emergency Communications: Packet radio is being used in many emergency services. Whether packet is used to pass a message accurately and in large quantities or to handle messages passed by the National Traffic System, it can provide an important function like any other amateur mode when used correctly. A new application called APRS combines GPS (Global Positioning Satellites) with packet radio to allow a master station to plot on their computer the location of all other stations in the field. The purpose is to coordinate the exact position of weather spotters or searchers, without having to waste radio time informing the control station of their locations. Recently, amateurs in Oklahoma have been distributing Doppler Radar images via the packet network. The small weather image file takes but a few minutes to retrieve and display. This helps those amateurs outside of the local ATV coverage to get an accurate weather picture from the Doppler Radar.

Networking: Since amateurs use radios to transmit their data, their range of communications is limited to approximately line of sight. An average packet station talks in a radius of about 10-30 miles. Packet Networks allow amateurs to widen the area of communications past their line of sight, by having a series of packet stations linked by radio that can be used to get their packet messages to where ever the network goes. Much like the telephone system, networks provide long distance service outside the local area. There are a number of amateur networks which allow amateurs to travel from one area to another. Network types include: Net/Rom, TCP/IP, TexNet, G8BPQ, ROSE, KaNodes, and many more.

These networks are typically built by a local or regional group that allows packet operators to get outside of their area. Amateurs get hooked on building and maintaining such networks, just like some amateurs operate DX or handle emergency communications. The type of network you use locally will depend on your area. Much depends on the network philosophy the local group has chosen when developing their network.

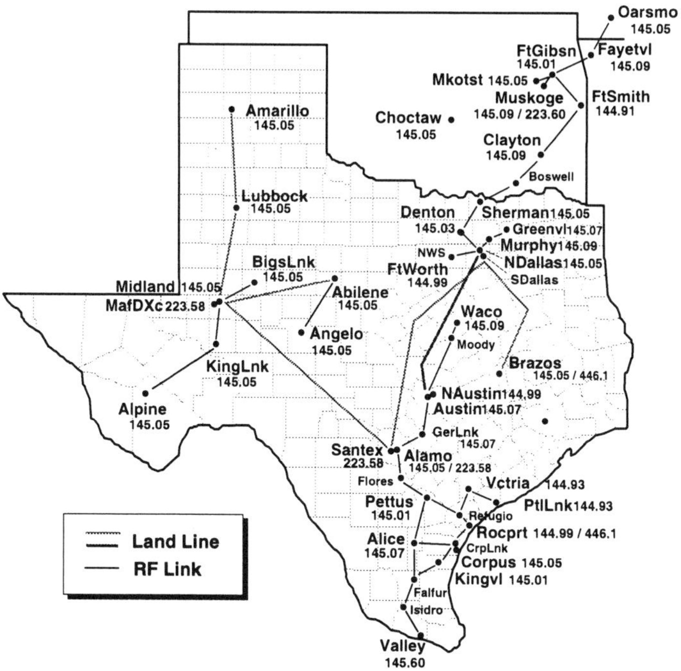

Satellite Communications: Many of the amateur radio satellites in orbit contain computer systems that provide packet capability. Most packet satellites provide BBS-like functions for messages to be passed to anywhere in the world within 24 hours. Several contain CCD cameras, which allow amateurs to download images of the earth and some allow users to retrieve data from the onboard experiments. Most satellites use AX.25 with special software developed for satellite communications. DOVE, Digital Orbit Voice Encoder, can be received with any normal VHF/FM 2-meter packet station, but most of the packet satellites use SSB and require more complex equipment in order to operate them. Just something else to spend your amateur dollars on.

Conclusion: These are just some of the things you can do with packet radio. Once you find something that you can do with packet radio, then you have a reason to purchase the equipment necessary to get on the air. A good place to start is to find a friend who uses packet and go visit. See what your local area has to offer. As already stated, packet radio changes every 50-mile radius. What is being done where I operate is probably slightly different than what you can do where you live.

Finding a local Elmer is an important first step.

References:

Finke, C. R. (Ed.) (1992, February 15). TPRS Quarterly Report. Texas Packet Radio Society, Inc.

Jones, G., G. Knezek, M. Hata. (1992). Packet Radio Prospects for Educational Data Communications. Proceedings of the Ninth International Conference on Technology in Education, 1, 218-219. Paris, France.

Lucas, Larry, Greg Jones, David Moore. (1992) An Educator's Alternative to Costly Telecommunications. Texas Center for Educational Technology, University of North Texas.

Introduction to Packet Radio

Greg Jones, WD5IVD

• What is packet radio?

Packet radio is a particular digital mode of Amateur Radio ("Ham" Radio) communications which corresponds to computer telecommunications. The telephone modem is replaced by a "magic" box called a terminal node controller (TNC), the telephone is replaced by an amateur radio transceiver, and the phone system is replaced by the "free" amateur radio waves. Packet radio takes any data stream sent from a computer and sends that via radio to another amateur radio station similarly equipped. Packet radio is so named because it sends the data in small bursts, or packets.

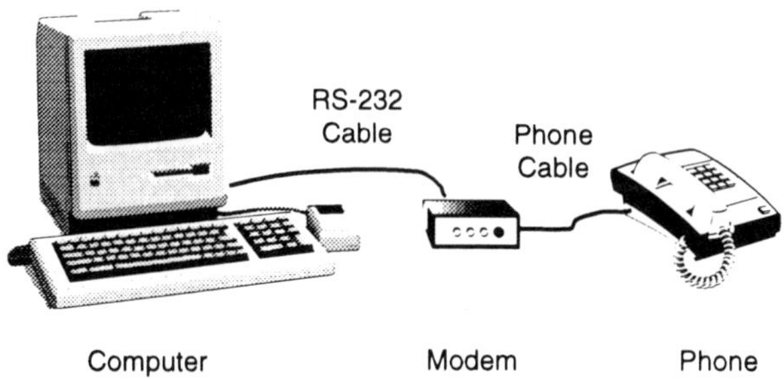

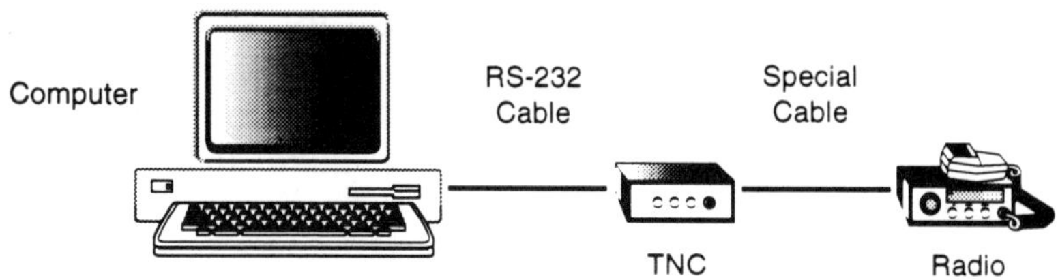

• What is the history of packet radio ?

Data packet technology was developed in the mid-1960's and was put into practical application in the ARPANET which was established in 1969. Initiated in 1970, the ALOHANET, based at the University of Hawaii, was the first large-scale packet radio project. Amateur packet radio began in Montreal, Canada in 1978, the first transmission occurring on May 31st by the Vancouver Amateur Digital Communication Group (VADCG).

The current TNC standard grew from a discussion in October of 1981 at a meeting of the Tucson Chapter of the IEEE Computer Society. A week later, six of the attendees gathered and discussed the feasibility of developing a TNC that would be available to amateurs at a modest cost. The Tucson Amateur Packet Radio Corporation (TAPR) formed from this project. On June 26th 1982, Lyle Johnson, WA7GXD, and Den Connors, KD2S, initiated a packet contact with the first TAPR unit. The project progressed from these first prototype units to the TNC-1 and then finally to the TNC-2 which is now the basis for most packet operations worldwide.

• Why packet over other modes?

Packet has three great advantages over other digital modes: transparency, error correction, and automatic control.

The operation of a packet station is transparent to the end user; connect to the other station, type in your message, and it is sent automatically. The terminal Node Controller (TNC) automatically divides the message into packets, keys the transmitter, and then sends the packets. While receiving packets, the TNC automatically decodes, checks for errors, and displays the received messages. Packet radio provides error free communications because of built-in error detection schemes. If a packet is received, it is checked for errors and will be displayed only if it is correct. In addition, any packet TNC can be used as a packet relay station, sometimes called a digipeater. This allows for greater range by stringing several packet stations together.

Users can connect to their friends' TNCs at any time they wish, to see if they are at home. Some TNCs even have Personal BBSs (sometimes called mailboxes) so other amateurs can leave messages for them when they are not at home. Another advantage of packet over other modes is the ability for many users to be able to use the same frequency channel simultaneously.

• What elements make up a packet station?

Figure 1 shows an illustration of a typical station setup with a schematic diagram of a station wiring.

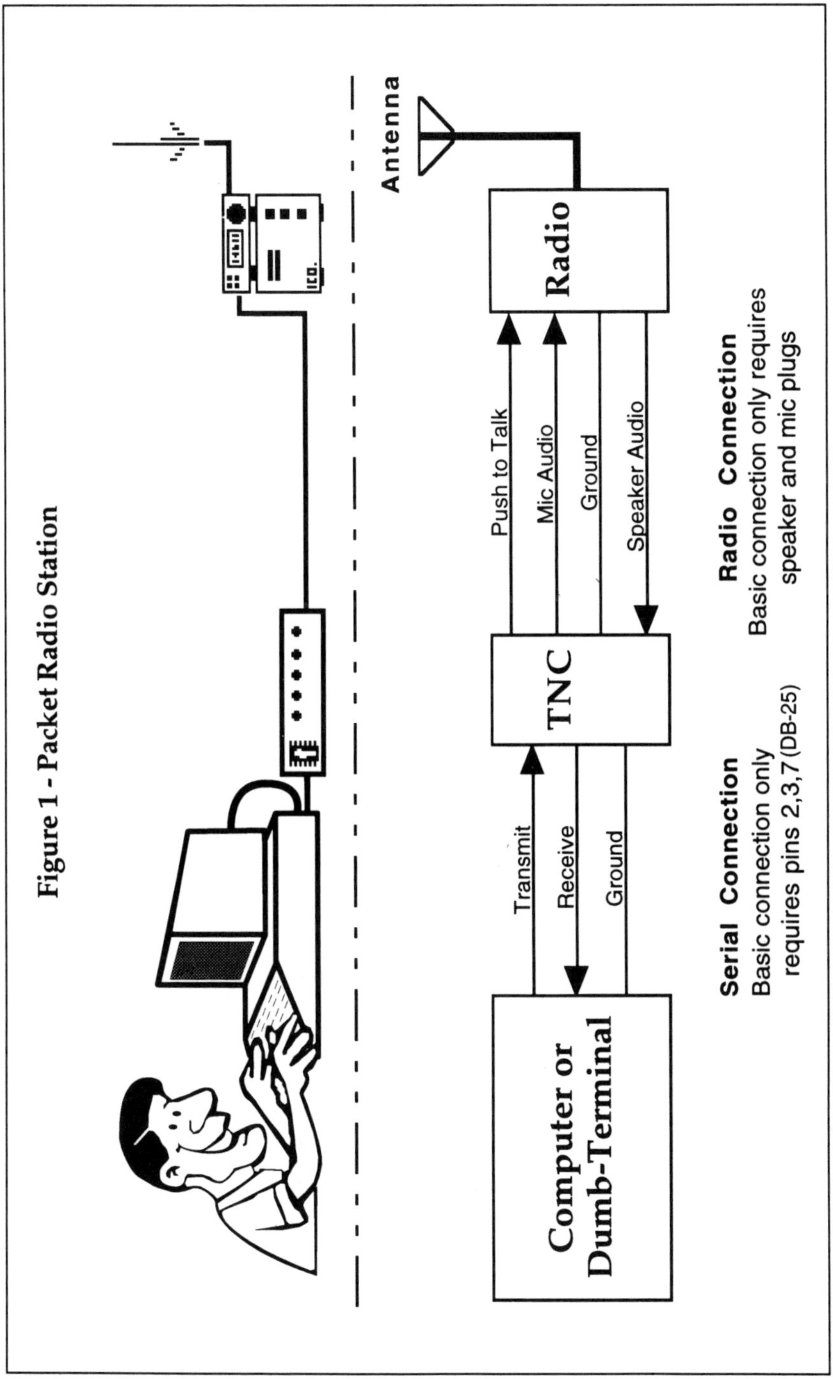

Figure 1 - Packet Radio Station

TNC (terminal Node Controller): A TNC contains a modem, a computer processor (CPU), and the associated circuitry required to convert communications between your computer (RS-232) and the packet radio protocol. A TNC assembles a packet from data received from the computer, computes an error check (CRC) for the packet, modulates it into audio frequencies, and puts out signals to transmit the packet over the connected radio. It also reverses the process, translating the audio that the connected radio receives into a byte stream that is then sent to the computer.

TNC Block Diagram and Description

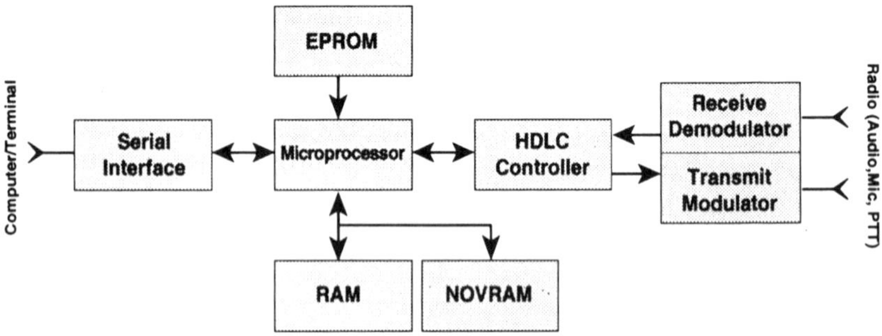

Serial Interface

- Information from your Computer feeds into the TNC via the Serial Interface.
- Information received by the TNC is sent to the computer via the serial interface.
- The speed of the computer to serial port connection is independent from the speed the modem operates at.
- Most TNC allow 1200 or 9600 bps communications.

EPROM (Erasable Programmable Read Only Memory)

- The EPROM stores the program which tells the TNC how to work.

NOVRAM (Nonvolatile Random Access Memory)

- The NOVRAM stores all the various user options (Callsign, TX Delay, Retry, etc).
- NOVRAM keeps information when the TNC is turned off.

RAM (Random Access Memory)

- The RAM is used by the microprocessor to stores information while the TNC operates.
- Anything stored in RAM when the TNC is powered off is lost.
- NOVRAM is used to store important information when the TNC is turned off.

HDLC (High-Level Data Link Control)

- The HDLC is used to turn the stream of information sent from the microprocessor to be transmitted into a stream of data (frames) for the modem to transmit.
- When the HDLC gets data from the receive demodulator, it turns the information into data the microprocessor can use.

Modem (Modulator/Demodulator)

- The modem is used to transmit and receive information to and from other TNCs.

Most amateurs currently use 1200 bps (bits per second) for local VHF and UHF packet, and 300 bps for longer distance, lower bandwidth HF communication. Higher speeds are available for use in the VHF, UHF, and especially microwave region. Most require special (not plug-and-play) hardware and drivers; 9600 baud plug-and-play TNCs and radios are new arrivals to packet.

Computer or Terminal: This is the user interface. A computer running a terminal emulator program, a packet-specific program, or just a dumb terminal can be used. For computers, almost any phone modem communications program (i.e. Procomm+, Bitcom, X-Talk) can be adapted for packet use, but there are also customized packet radio programs available. A dumb terminal, while possibly the cheapest option, does have several limitations. Most dumb terminals do not allow you to scroll backwards, store information, or upload or download files.

A radio: For 1200/2400 bps UHF/VHF packet, commonly available narrow band FM voice radios are used. For HF packet, 300 BPS data is used over single side band (SSB) modulation. For high speed packet (starting at 9600 bps), special radios or modified FM radios must be used. 1200 bps AFSK TNCs used on 2-meters (144-148Mhz) is the most commonly found packet radio.

• What is the distance limitation for packet radio ?

Since packet radio is most commonly used at the higher radio frequencies (VHF), the range of the transmission is somewhat limited. Generally, transmission range is limited to "unobstructed line-of-sight" plus approximately 10-15%. The transmission range is influenced by the transmitter power and the type and location of the antenna, as well as the actual frequency used and the length of the antenna feed line (the cable connecting the radio to the antenna). Another factor influencing the transmission range is the existence of obstructions (hills, groups of buildings, etc). Thus, for two-meter packet (144 - 148Mhz), the range could be 10 to 100 miles, depending on the specific combination of the variables mentioned above.

• What do you mean we can all use the same channel?

Packet radio, unlike voice communications, can support multiple conversations on the same frequency at the same time. This does not mean that interference does not occur when two stations transmit at the same time, known as a collision. What 'same time' means in this sense is that multiple conversations are possible in a managed, time shared fashion. Conversations occur during the times when the other conversations are not using the channel. Packet radio uses a protocol called AX.25 to accomplish this shared channel.

AX.25 specifies channel access (ability to transmit on the channel) to be handled by CSMA (Carrier Sense Multiple Access). If you need to transmit, your TNC monitors the channel to see if someone else is transmitting. If no one else is transmitting, then the TNC keys up the radio, and sends its packet. All the other stations hear the packet and do not transmit until you are done. Unfortunately, two stations could accidentally transmit at the same time. This is called a collision. If a collision occurs, neither TNC will receive a reply back from the last packet it sent. Each TNC will wait a random amount of time and then retransmit the packet. In actuality, a more complex scheme is used to determine when the TNC transmits. See the "AX.25 Protocol Specification" for more information (ARRL, 1988).

• What is AX.25?

AX.25 (Amateur X.25) is the communications protocol used for packet radio. A protocol is a standard for two computer systems to communicate with each other, somewhat analogous to using a business format when writing a business letter. AX.25 was developed in the 1970's and based on the wired network protocol X.25. Because of the difference in the transport medium (radios vs wires) and because of different addressing schemes, X.25 was modified to suit amateur radio's needs. AX.25 includes a digipeater field to allow other stations to automatically repeat packets to extend the range of transmitters. One advantage of AX.25 is that every packet sent contains the sender's and recipient's amateur radio callsign, thus providing station identification with every transmission.

• Networking and special packet protocols

This is a sample of some of the more popular networking schemes available today. By far, there are more customized networking schemes used than listed. Consult your local packet network guru for specific network information.

• Are there any other protocols in use other than AX.25?

AX.25 is considered the defacto standard protocol for amateur radio use and is even recognized by many countries as a legal operation mode. However, there are other standards. TCP/IP is used in some areas for amateur radio. Also, some networking protocols use packet formats other than AX.25. Often, special packet radio protocols are encapsulated within AX.25 packet frames. This is done to insure compliance with regulations requiring packet radio transmissions to be in the form of AX.25. However, details of AX.25 encapsulation rules vary from country to country.

• Networking Schemes

What are some of those other networking schemes?
During the early days of amateur packet radio, it became apparent that a packet network was needed. To this end, the following packet network schemes were created.

Digipeaters: The first networking scheme with packet radio was Digipeaters. Digipeaters would simply look at a packet, and if its call was in the digipeater field, would resend the packet. Digipeaters allow the extension of range of a transmitter by retransmitting any packets addressed to the digipeater. This scheme worked well when only a few people were on the radio channel. However, as packet became more popular, digipeaters soon were clogging up the airwaves with traffic being repeated over long distances. Also, if a packet got lost by one of the digipeaters, the originator station would have to retransmit the entire packet again, forcing even more congestion.

KA-Nodes: Kantronics improved on the digipeater slightly and created KA-Nodes. As with digipeaters, KA-Nodes simply repeat AX.25 frames. However, a KA-Node acknowledges every transmission at each link (node) instead of over the entire route. Therefore, instead of an end-to-end acknowledgment, KA-Nodes allow for more reliable connections with fewer timeouts, because acknowledgments are only carried on one link. KA-Nodes therefore are more reliable than digipeaters, but are not a true network. It is similar to having to wire your own telephone network to make a phone call.

NET/ROM: NET/ROM was one of the first networking schemes to try to address the problems with digipeaters. A user connects to a NET/ROM station as if connecting to any other packet station. From there, he can issue commands to instruct the station to connect to another user locally or connect to another NET/ROM station. This connect-then-connect again means that to a user's

TNC, you are connected to a local station only and its transmissions do not have to be digipeated over the entire network and risk losing packets. This local connection proved to be more reliable.

NET/ROM doesn't use all of the AX.25 protocol. Instead, it uses special AX.25 packets called Unnumbered Information (UI) packets and then puts its own special protocol on top of AX.25. This is again used to increase efficiency of its transmissions. NET/ROM nodes, at regular intervals, transmit to other nodes their current list of known nodes. This is good because as new nodes come on-line, they are automatically integrated in the network. However, if band conditions such as ducting occur, ordinarily unreachable nodes can be entered into node lists. This causes the NET/ROM routing software to choose routes to distant nodes that are normally impossible. This problem requires users to develop a route to a distant node manually defining each hop instead of using the automatic routing feature.

NET/ROM is a commercial firmware (software put on a chip) program that is used as a replacement ROM in TAPR type TNCs. Other programs are available to emulate NET/ROM. Among them are TheNet, G8BPQ node switch, MSYS, and some versions of NET.

ROSE: ROSE is another networking protocol derived from X.25. Each ROSE node has a static list of the nodes it can reach. For a user to use a ROSE switch, he issues a connect with the destination station and in the digipeater field places the call of the local ROSE switch and the distant ROSE switch the destination station can hear. Other than that, the network is completely transparent to the user.

ROSE's use of static routing tables ensures that ROSE nodes don't attempt to route packets through links that aren't reliably reachable, as NET/ROM nodes often do. However, ROSE suffers from the inability to automatically update its routing tables as new nodes come on-line. The operators must manually update the routing tables, which is why ROSE networks require more maintenance.

TCP/IP: TCP/IP stands for Transmission Control Protocol/Internet Protocol. TCP/IP is commonly used over the Internet wired computer network. The TCP/IP suite contains different transmission facilities such as FTP (File Transfer Protocol), SMTP

(Simple Mail Transport Protocol), Telnet (Remote terminal protocol), and NNTP (Net News Transfer Protocol) The KA9Q NOS program (also called NET) is the most commonly used version of TCP/IP in packet radio. NOS originally was written for the PC compatible. However, NOS has been ported to many different computers such as the Amiga, Macintosh, Unix, and others. Smaller computers like the Commodore 64 and the Timex-Sinclar do not currently have versions of NOS available. TCP/IP based amateur networks are becoming more common each day.

TexNet: TexNet is a 3-port switch designed to create a 9600 baud backbone with 2 local access channels. The TexNet network provides transparent network access to the user. The user simply accesses his/her local TexNet node and then either connects to a user at another node or accesses various system services. TexNet provides the stability of fixed routing, while allowing new nodes to be automatically brought into the network.

BBS Message Transfer: Many of the BBS programs used in packet radio allow for mail and bulletins to be transferred over the packet radio networks. The BBSs use a special forwarding protocol developed originally by Hank Oredsen, W0RLI. Besides full service BBSs, many TNC makers have developed Personal BBS software to allow full service BBSs to forward mail directly to the amateur's TNC. This allows operators to receive packet mail at night and avoid tying up the network during busy hours.

References:

Finke, C. R. (Ed.) (1992, February 15). TPRS Quarterly Report. Texas Packet Radio Society, Inc.

Jones, G., G. Knezek, M. Hata. (1992). Packet Radio Prospects for Educational Data Communications. Proceedings of the Ninth International Conference on Technology in Education, 1, 218-219. Paris, France.

Lucas, Larry, Greg Jones, David Moore. (1992) An Educator's Alternative to Costly Telecommunications. Texas Center for Educational Technology, Univ. of North Texas.

Steve Watt, KD6GGD, steve@wattres.SJ.CA.US. (1993). Frequently Asked Questions from the listing in the rec.radio.amateur.packet newsgroup for the USENET network. Version 1.11.

Tucson Amateur Packet Radio Corporation. terminal Node Controller Manual, Firmware Release 1.1.8, Tucson, AZ: Author.

This article reprinted by premission of the Amateur Radio Relay League. Aritlce appeard in the 1993 September QST. ARRL (203) 666-1541

Go Digital!

The HF digital modes are booming. There's no better time than now to join the fun.

By Kirk Kleinschmidt, NT0Z, Assistant Managing Editor

When most hams think of computers and Amateur Radio, packet comes to mind. And why not? During last 10 years packet has become the most popular digital communication mode. More than 50,000 packeteers couldn't be wrong!

Packet is terrific for local networks on VHF and UHF, but it's not well suited for HF work. When it comes to HF digital hamming, the favorite modes are RTTY, AMTOR, PacTOR, CLOVER, and G-TOR. This collection looks like alphabet soup, doesn't it? Let's dispel some of the mystery by taking a brief look at the hardware and software required to get started in these fascinating modes, and what you can do with them on the air.

The Transceiver

You'll find that most modern HF transceivers easily accommodate the digital modes. If you're considering AMTOR, PacTOR, CLOVER, or G-TOR, make sure your rig can switch from transmit to receive very rapidly. For example, AMTOR requires a transmit/ receive turnaround time of 30 milliseconds or less for long-distance work.. QST Product Reviews and some manufacturer's ads list these specs.

If you're interested in RTTY, you'll need a rig that can tolerate high-duty-cycle transmissions for extended periods of time. (CW and SSB are low-duty-cycle modes because they cause the transmitter to produce full output for only brief intervals.) Most modern rigs are rated for high-duty-cycle use, but some must be operated at reduced output.

Older rigs are often less tolerant of high-speed switching and high-duty-cycle transmissions. AMTOR, PacTOR, CLOVER and G-TOR are impossible on many vintage radios. Most older SSB transmitters and transceivers should be RTTY compatible — although you'll have to be careful not to exceed their specifications for continuous output power!

Here's an easy rule of thumb for selecting an HF transceiver for digital applications: If you intend to operate any of the modes that require rapid transmit/receive switching, buy a rig that was manufactured within the last five to ten years. If your only interest is RTTY, almost any good-quality radio made within the last 20 years will do the job.

The Computer

With all due respect to owners of Apple, Commodore, Atari and other computer brands, it's fair to say that IBM PC compatible computers have become the de facto standard in Amateur Radio. This doesn't mean that other models are unusable for digital communications, though. (The only exception is CLOVER which requires an IBM-PC or compatible.) If you can get a terminal-emulation program for your computer, you're halfway there.

Virtually any personal computer is sufficient to get you off and running. There are a few factors to consider before making your choice, however:

I/O (input/output): Does the computer have an RS-232C serial port? Most digital interfaces need one.

Software: Is there a terminal-emulation program available for the computer?

Memory: How much random-access memory (RAM) is available? When it comes to RAM, the more the merrier.

Data storage: Does the computer offer some form of reliable data storage (preferably floppy disks or a hard disk)?

Interference: Does the computer bombard your receiver with RF of its own making? How does it respond to your radio signals? Interference can be a major headache!

Support: If your computer suddenly becomes demented, is there anyone on the planet who remembers how to fix it?

The computer that's best for you will ultimately depend on your budget and your goals. If you're a little unsure about digital communications, start small and work your way up.

The Analog/Digital Interface

Getting your computer to talk to your radio is the job of the interface, which must accept digital data from the computer and translate it into audio signals (or on/off keying) for the transceiver, and vice-versa. Unless you're using a CLOVER card in an IBM-PC, you'll need an interface device. In this case you're really talking about modems—much like the ones used to link computers via telephone.

Multimode Communications Processors

As digital technology evolved, it wasn't long before someone said, "Why should I have to use one interface for packet and another for RTTY? Why can't they all be together in one box?"

Not only did manufacturers place packet and RTTY together in the same box, they added several other modes for good measure. The result was the multimode communications processor (or MCP)—probably one of the most popular pieces of equipment for the digital ham today.

Like packet TNCs, MCPs incorporate microprocessor design and internal memory. By issuing a single command from the computer keyboard, you can switch the MCP from packet (HF or VHF), to AMTOR, to RTTY or whatever. Some models also add fax, ASCII, NAVTEX and SSTV to their list of modes. MCPs have brought a smorgasbord of operating modes within the reach of average hams. Their sheer convenience is addictive!

Digital Signal Processing

Now that you've been introduced to MCPs, it's time to tell you about an exciting new technology called digital signal processing or DSP—an entirely digital approach to decoding and encoding signals for various modes. Its main advantage is its flexibility.

For example, you can add a new digital mode to a standard MCP by replacing an IC chip that has been programmed with new software. This is fine as long as the unit can support the new mode. But what if you want the processor to perform a task that's beyond the capability of its existing hardware?

With a DSP processor, you can add virtually any mode to the unit by simply adding new software. Unlike standard communications processors, DSP units use software to directly encode and decode signals without depending on specialized hardware.

RTTY

Technically speaking, the term RTTY encompasses many forms of amateur digital communications, including AMTOR and ASCII. In common use, however, RTTY refers to the use of the 5-bit International Telegraph Alphabet Number 2 (ITA2), also known as Baudot (pronounced Baw-dough). Although communications data rates for landline or radio modems are properly expressed in bits per second, you'll often see them referred to as so many "bauds." Telephone modems typically run at 2400 or 9600 bauds; most VHF packet runs at 1200 bauds; AMTOR chirps along at 100 bauds and most amateur RTTY warbles along at an easy-to-type pace of 45.45 bauds. Further discussion of the technical aspects of RTTY are beyond the scope of this article. For more technical information, check out The ARRL Handbook or the ARRL Operating Manual.

Unlike AMTOR, which we'll get to in a minute, making RTTY contacts is a lot like making Morse code QSOs. Because no error-correction schemes are used, signal strength and propagation are important. If a RTTY signal fades out for a second, your RTTY terminal "loses copy."

To make a contact, one station usually calls CQ by turning on his transmitter and typing "CQ CQ CQ DE NT0Z NT0Z NT0Z K K K," or something similar, usually repeated two or three times (computerized terminals and their programmable buffers make short work of this).

A station answering the CQ might reply with a three-by-three reply (sending the CQ'ers call sign three times followed by his call sign three times), which is sometimes preceded by a short string of "RYs." Because of their Baudot code patterns, an "endless" string of the letters R and Y produces an unmistakable sound. The RYs make it easy for the receiving station to exactly tune in the other station before information is exchanged—sort of like priming the pump. (Don't overdo the RYs, however . . .)

From then on, an RTTY QSO is a lot like a Morse code QSO. In fact, many of the same procedural signs are used.

If the hustle and bustle of everyday hamming is getting you down, RTTY may be the answer. RTTY ops are known for being extra friendly, and "hitting the keys" makes for an interesting, almost nostalgic, ragchew medium. Ragchewing is a popular RTTY pastime, but there' something for everyone, from intense contest action, to traffic handling, to awards chasing and DXing. (With low-cost RTTY technology catching up to the rest of the world, the RTTY subbands are bursting with DX stations. Often, they're much easier to work than their SSB and CW counterparts.) One look at the DXing newsletters and you'll see that even DXpeditions to rare spots around the globe equip at least one RTTY station. If you've "worked 'em all" an SSB or CW, why not give RTTY a try?

Regardless of your interests, even a fully computerized RTTY setup seems to take you back to a bygone era when mechanical teleprinters churned out copy on long rolls of gritty yellow paper. You can almost smell the oil and the machinery . . .

AMTOR

If RTTY is known for its nostalgic appeal, then AMTOR, a modern error-correcting cousin of SITOR, brings amateur HF digital communications into the computer age.

Instead of sending and receiving steady streams of data, an AMTOR station sends short data bursts back and forth, and will keep repeating the same burst until the receiving station has acknowledged receipt. This results in AMTOR's (and SITOR's) characteristic "chirp-chirp-chirp" sound, which is hard to miss on the bands.

If signals fade momentarily, the AMTOR stations chirp away until the data has been correctly sent and received. This eliminates some of the "nostalgic" feel, but ensures nearly 100% copy, which is important in traffic handling or transferring computer or text files. Ragchewing via AMTOR is a lot like hooking up via VHF packet radio.

All of the previously mentioned RTTY activities take place on AMTOR, but the newer mode is more suitable for certain types of "remote access" activities such as AMTOR "mailboxes," which work a lot like landline BBSs. You can leave mail for other AMTOR users, and when they "check in," their mail is waiting to be picked up. Text files and other data files are also available from the mailbox.

Probably the most popular AMTOR mailbox system is a global network of APLink stations. APLink stations can network with VHF packet networks, for seamless traffic handling and emergency communications. As an example of how effective AMTOR can be, during the Gulf War, hundreds of messages were clandestinely relayed from occupied Kuwait by only a handful of AMTOR/APLink operators hidden within the borders.

PacTOR

PacTOR made its debut in the early 90s and it's been growing rapidly ever since. PacTOR has become so popular, it has almost replaced AMTOR as the most active HF digital communications mode.

Like AMTOR and packet, PacTOR sends error-free information by using a handshaking system. When the data is received intact, the receiving station sends an ACK signal (for acknowledgment). If the data contains errors, a NAK is sent (for nonacknowledgment). AMTOR and PacTOR sound similar when you hear them, but PacTOR is the mode with the extended chirps.

The key to PacTOR's magic is a feature known as memory ARQ. Each time your MCP receives an incomplete data burst (one with missing data), the bits of data that made it through intact are stored temporarily in memory. When the data burst is repeated, memory ARQ looks for the missing data and attempts to fill the gaps. In most cases, memory ARQ can assemble error-free information with only one or two repeat transmissions. The fewer the repeats, the faster your communication will be! (PacTOR is much faster than AMTOR under most conditions.)

PacTOR also adjusts itself automatically to the band conditions. If band conditions are good, PacTOR will run at 200 bit/s (bits per second). If the band starts to go, it adjusts itself down to 100 bit/s.

With memory ARQ and automatically adjustable data rates, PacTOR offers very good performance. PacTOR also supports the complete ASCII character set. This means that you can send every letter of the alphabet (in upper or lower case) as well as punctuation marks and special symbols. (AMTOR supports only a limited character set.) It's also possible to swap binary files on PacTOR.

Most APLink systems now include PacTOR. You can connect and pass traffic using either PacTOR or AMTOR as you prefer. There are also plenty of PacTOR BBSs on the air as well as hams eager to enjoy live keyboard-to-keyboard chats.

CLOVER

Of all the Amateur Radio HF digital communication modes, CLOVER offers the best overall performance. Using a complicated modulation scheme, automatic output power adjustment and other advanced features, CLOVER is able to maintain communication under the worst conditions. CLOVER is also very efficient. In terms of throughput, CLOVER is the most efficient HF digital mode available to hams today.

Although CLOVER appeared at nearly the same time as PacTOR, it hasn't enjoyed PacTOR's success. This has been mainly due to hardware requirements and cost. This situation is changing, as you'll see in a moment.

To operate CLOVER you must purchase a PCI-4000 card from HAL Communications. This card installs in the expansion slots of an IBM-PC computer (286 or or higher). The connections to your SSB transceiver are simple. One cable goes to your microphone jack and another connects to your external speaker jack or auxiliary audio output. Almost any modern SSB transceiver will work with CLOVER—if it is very stable.

Like AMTOR and PacTOR, CLOVER uses a handshaking system to exchange information. In the case of CLOVER, however, stations swap information about signal conditions and power output levels in addition to the files or other data they're attempting to send. The length of CLOVER bursts can vary, but they're usually longer than AMTOR or PacTOR.

When CLOVER cards were first available, the cost was about $1000 per card. A CLOVER card could operate only in CLOVER, no other modes were available. Prices of CLOVER cards have fallen to around $700 or $800 and they are now multimode units, offering RTTY, AMTOR and PacTOR in addition to CLOVER. With the reduced cost and multimode features, you may hear more CLOVER stations on the air soon. At the present time CLOVER is being used primarily as a means to transfer large amounts of data between bulletin board systems.

G-TOR

G-TOR is the newest Amateur Radio digital mode in active use. Pioneered by Kantronics Corporation, G-TOR was released early this year as a new feature of their popular KAM Plus MCP.

G-TOR is an acronym for Golay-coded Teleprinting Over Radio. Golay coding is the error-correction system created by M.J.E. Golay and used by the Voyager spacecraft. Sending billions of bytes of data across the Solar System required a scheme to ensure that the information could be recovered despite errors caused by interference, noise, and so on. It seemed like a perfect choice for overcoming some of the difficulties of digital communication on the HF bands.

Operating G-TOR is essentially the same as operating AMTOR or PacTOR. G-TOR signals sound almost identical to PacTOR when you hear them on the air. (If you listen carefully, the G-TOR bursts are longer than PacTOR bursts.) Like PacTOR, G-TOR supports the complete ASCII character set in addition to the ability to transfer binary files.

The main advantage of G-TOR over AMTOR and PacTOR is speed. G-TOR is capable of transferring data at rates nearly triple those of PacTOR—even under terrible band conditions. G-TOR even approaches the efficiency of CLOVER in certain situations.

When this article went you press, you needed a Kantronics KAM Plus (or an upgraded KAM) to operate G-TOR. Other MCP manufacturers have also expressed interest in adding G-TOR to their products. The popularity of G-TOR is on the rise, but it's too early to tell whether it will give PacTOR any serious competition.

Other Modes

Also available to today's digital hams are modes like ASCII (the upper- and lower-case "alphabet" used by your computer), slow-scan television (SSTV) and facsimile. Sending fax pictures on HF used to be reserved for those who happened to own expensive and bulky commercial or maritime fax transceivers. Because today's MCPs offer black-and-white fax capabilities, sending amateur fax pictures is becoming more popular. Who knows, fax may become the next mode to explode with activity!

Table 1

Common HF Digital Frequencies (in kHz)

1800-1840
3605-3645
7080-7100
10,140-10,150
14,070-14099.5
18,100-18,110
21,070-21,100
24,920-24,930
28,070-28,150

Do's and Don'ts: Rules and Regulations

Steve Watt, KD6GGD, steve@wattres.SJ.CA.US. Frequently Asked Questions Version 1.11 from rec.radio.amateur.packet.

NOTE: These regulations apply only to amateurs regulated by the FCC (United States), but often are similar to regulations in other countries.

Can I set up a TNC at home and one at work so I can check my Electronic mail via packet?
Amateur radio rules **prohibit** any business. Since you could have mail from your boss or something business related over Internet, that constitutes business activity and is specifically prohibited. Profanity can also be a complication. Since you have no control over the language used in E-mail, proper filtering is required — thus every message must be hand filtered. You have to be VERY careful about what mail you get.

I would like to set up a packet radio gateway between a land line computer network and the packet network. Is this possible?
Yes, and there are several such gateways in use, but they must be managed with caution. Electronic mail may be passed FROM the packet network INTO the land line network without intervention. However, mail passed TO packet radio is considered third party traffic (the sender is not an amateur) and these messages must be hand filtered.

It's my license if I use packet radio illegally anyway, so what does anyone else care!?
Packet radio is one of the few NETWORKED systems in amateur radio. Sending packet BBS electronic mail, digipeating, and sharing the channel involves the licenses of MANY people. When in doubt, it is best to check with other amateurs about sending the message before it is sent.

I have some ideas on how to use packet radio in a new way, but I don't know if it is legal. Who could tell me if I can do it legally?
The worst thing you can do is talk to the FCC about such an issue. The FCC rules are written to be general enough to encompass but not restrict new radio activities. In the past, any non-thought-out requests sent to the FCC have meant a reduction of privileges. The best source for legal assistance is the American Radio Relay League (ARRL). Another good place for such conversations is over the various TAPR SIGs on Internet, or the amateur radio BBS network.

Buying Equipment: Terminal Node Controllers

Greg Jones, WD5IVD

Used to be a few years ago, I could count the number of TNCs available to choose on one hand. Made it downright simple to make a decision on which one to buy, since the choices were pretty straightforward. Now, there are lots of choices. A recent amateur equipment catalog had twenty TNCs on one page! To make a decision now, you have three main issues that determine what you want: Price, Functionality, and Configuration. Prioritize these and you probably have made your choice. The other thing that can help is to find another ham that has the equipment you are interested in, and then go take a look at it in operation. Sometimes seeing how something works can really help.

There are two types of TNCs. The traditional TNC (terminal Node Controller) that does nothing but AX.25 packet radio. There is not much difference between the packet only TNCs, except the different modem options they have: 300baud AFSK, 1200baud AFSK, 2400bps DPSK, 9600baud FSK, etc). Some of the the things you might look for is HF capability or a personal mailbox option. Your other choice is to purchase a multi-mode controller, which does a little of everything: AX.25 Packet, AMTOR, Morse, ASCII, RTTY, WEFAX, PACTOR, Fax, etc, etc, etc. There are several multi-mode controllers available and each have a little bit different functionality, but all do about the same basic functions. Functionality and price are interrelated with both types of units. Plain TNCs that do just AX.25 1200 baud AFSK packet are going to cost around $120. These are the de facto standard TNCs on 2 meter packet radio. Multi-mode controllers begin at $250 and go up depending on how many bells and whistles are provided. If you are wanting to just get on and play VHF packet only, then the 'TNC only' units are the best buy. VHF and UHF operations are typically at 1200 baud AFSK and these units allow you to to get on packet with an initial low investment. A multi-mode controller allows you to play with other modes besides packet, but has a higher initial cost. Many of the additional functions of the multi-mode controllers are for use on the HF bands, so have limited use on VHF/UHF. This initial higher cost might not be worth the investment if you do not operate HF frequencies.

Many beginners get the lower cost 'TNC only' unit to determine if they like digital communications and then later dedicate the unit for local VHF operations and then purchase a multi-mode controller to branch into new modes. This is a tried and true method. Examples of packet only units include: TAPR TNC kit, MFJ-1270 / 1274, DRSI DPK-2, Kantronics KPC-3, and AEA PK-96 / PK-88. Examples of multi-mode controllers include: MFJ-1278, Kantronics KAM, and AEA PK-232 / PK-900 / DSP-2232.

AEA PK-232 Multi-mode Dual Port Controller. Courtesy AEA.

Configuration deals with External versus Internal installation. TNCs are available as external units that stand alone and internal units that plug into an IBM PC/Clone computer. With an external unit, power, an interface cable to the radio, and a serial cable to your computer are needed. Examples of external units include: MFJ 1270, Kantronics KPC-3, DRSI DPK-2, and AEA PK-88. Internal units are simply PC plug-in cards that fit inside your computer; thus, they get their power and communications line from the computer's bus connection. This allows the unit to do without a power supply and box, making these units a little less expensive. Examples of these include: DRSI PCPA series, PKT, and AEA PCB-88. There are advantages to both configurations. Some hams prefer to have the equipment inside their computer so as not to take up space in the shack. Others prefer the convenience of being able to remove their TNC for use at another location without disturbing the computer. Configuration is probably the lowest priority of the three areas.

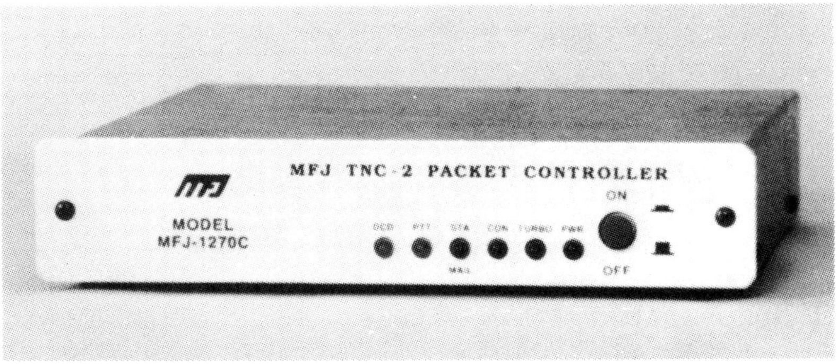

MFJ TNC-2 1270C Packet Controller. Courtesy MFJ.

One option I haven't discussed yet is the software resident TNC. With the software resident TNC, only the modem is hardware and the rest is done as software in the computer. Examples of software resident solutions include: BayCom, BayPac, and Poor Man's Packet. The advantages of this type of unit are that it is simple, small, and less expensive than the standalone TNC described above. The disadvantage is that software is available mostly for the IBM PC/Clones and for most users the software occupies the entire computer, thus not allowing you to use other software when you are doing packet. These units are often found in emergency operations due to their size and portability. If you plan to just operate packet infrequently, then the software resident TNC might be a good choice. TAPR has introduced a modem kit with software that allows RTTY, AMTOR, and PCTOR operations, a software resident solution to purchasing a multi-mode controller. It allows you to get on HF digital with a small initial investment to see if you like it.

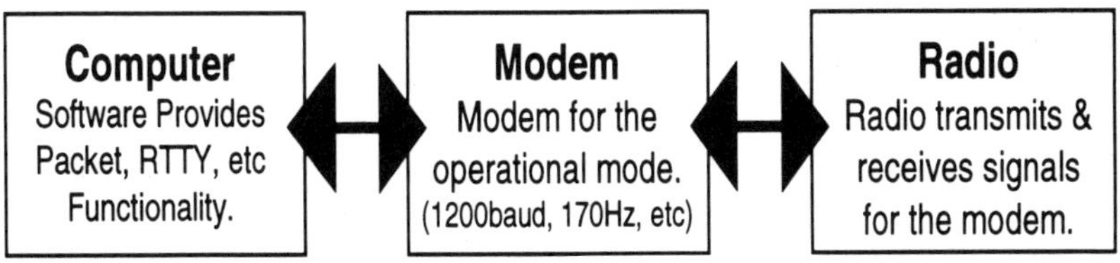

Diagram of a Software Resident System. The only hardware is the modem, since the computer is handling TNC/other processing functions.

Let's Review our choices. You should ask yourself:

1) What do you want to do with your computer and radio ?
2) How much money do you have ?
3) Is this a serious interest or something you want to experiment with ?
4) What is it you intend to access with the equipment ?

Once you answer these questions, you can probably determine the amount of investment you are willing to make and in what area. The equipment by itself does not make the fun. The resources you access with the equipment provides the enjoyment aspect to our hobby, but depends on the critical mass of equipment that others have. So, your choice will probably hinge on what locals are doing. If you are just beginning in packet radio, find someone in your local area and go take a look at what is available and what they are doing. You should be able to find several amateurs around doing different things; so go see what everyone is doing and pick something that looks like fun and get your equipment. If it does not look like it is increasing your fun factor — time to find something else to do. Purchasing similar equipment to that of someone you know is not a bad choice either, since then you can get local help and support when you are having problems. Having a local elmer can save numerous hours once you purchase your equipment. Many a potential packeteer has been lost to those screens of meaningless data scrolling past on the computer. Find an elmer, write down the few things it takes to get you on the local resources and then learn the rest later. Have fun and then work on increasing your knowledge. When you are having fun, learning comes much easier. Purchasing the equipment is the first step down that road.

References:

Finke, C. R. (Ed.) (1992, February 15). TPRS Quarterly Report. Texas Packet Radio Society, Inc.

Jones, G. (Ed.) (1993). Packet Radio General Information. Tucson Amateur Packet Radio Corp.

Jones, G., G. Knezek, M. Hata. (1992). Packet Radio Prospects for Educational Data Communications. Proceedings of the Ninth International Conference on Technology in Education, 1, 218-219. Paris, France.

Lucas, Larry, Greg Jones, David Moore. (1992) An Educator's Alternative to Costly Telecommunications. Texas Center for Educational Technology, University of North Texas.

Buying Equipment: Antennas for Packet

Jim Neely, WA5LHS

What follows may sound like heresy, but it makes sense to me and I thought I'd share it with others. Since the days of spark gap, we have been trying to improve our antennas to squeeze every bit of RF from our transmitters to the ether. For VHF packet, this might be counter productive.

I have prided myself on the quality of VHF antennas I have used since I first ran 2 meter AM on MARS nets during the Viet Nam war. It took some doing to be able to talk reliably on 150 mile paths. With some brute strength and a little luck, you could probably make yourself heard at that distance, but it doesn't do much good if you can't hear the station you are talking to. Lots of aluminum in the air and the absence of clouds with rocks in them makes the feat possible. Given the AX.25 protocol, do we really want to be able to hear every packet signal for 150 miles? The TNC-2 and its derivatives conform to the AX.25 protocol and refrain from transmitting if another signal is heard. This feature is designed into the protocol to eliminate collisions. If your high performance antenna at 90 feet hears every packet signal for 150 miles, it will never transmit.

This brings me to the suggestion that since we are using networks to communicate with each other, we only need enough power and antenna to communicate with the closest network node. That means if you can talk to the network node you usually use with a ground plane and 10 watts, that in my opinion, is what you should use, not 100 watts and a beam at 90 feet. The rules and regulations we live by state that we should use only the amount of power necessary to communicate. This brings up another point. If you can hit the closest digipeater or network node with your Handi-Talki, is that what you should use? I have suggested to newcomers to packet that they should get an old crystal-controlled, inexpensive radio rather than try to use their handheld radios for packet. You may be able to hear the station and have enough power to talk to it, but you are creating a hidden transmitter and most likely causing collisions.

I think that the use of only enough antenna to make the necessary communication and enough power not to be a hidden transmitter are just common sense and it makes your life much simpler. I hope the ideas presented here give rise to some thought and they might even explain some of our problems and the solutions to them.

Buying Equipment: Radios for Packet

Greg Jones, WD5IVD

As Jim mentions above, he recommends finding a crystal-controlled radio. This is not a bad recommendation, since frequency agility is not typically a high priority when looking for a radio to use on Packet Radio. These radios provide decent transmit power (10-15 watts) and can be found for less than $100 at hamfests. Combine 10 watts with an outside antenna and you have a good packet radio station that can be heard generally by everyone in the local area.

Why is it important to be heard ? Because of something called "Hidden Terminal". A "Hidden Terminal" is a packet radio station that is not heard by all of the stations on a given radio channel. You may become the "Hidden Terminal" whenever other packeteers on the same radio channel cannot hear your frames. Whenever you send a packet, any station that "hears" your packet frames should have his/her TNC's DCD (Data Carrier Detect) respond to your packet. The DCD signal inside the normal packet controller prevents the TNC from keying the radio and sending packet frames until the channel is clear (avoiding transmitting on someone else). Without the benefit of your signal's resultant DCD holding off the other station's packets, the other station transmits directly over your packets. Whenever this occurs, the PBBS, network node, or other station hears both packets at the same time. The other station then cannot decode either station's packets. When this happens, both your station and the other stations send the same packets over again. Your TNC will retransmit the same packet frames over again until it exceeds its retry limit. Each time your TNC retransmits and the other stations on the channel can't hear you, they also retransmit! The result? Collision City! Stations time out or disconnect on retries and the channel turns to a colossal waste of of time. So a radio with low power (Handi-Talki) on an inside antenna is a good candidate to be a Hidden Terminal. The Commission's Rules (Part 97.313(a)) state "An amateur station must use the minimum transmitter power necessary to carry out the desired communications." This means that the packet station must use enough power to establish and maintain communications, especially when working a packet BBS station. The key word here is "maintain," as in keeping a successful packet link or connection. Since packet radio is a shared channel resource, it is important that good radio and antenna configuration be maintained in order to allow everyone equal access to those limited resources. The real question is — do you have enough signal strength to ensure that all stations within range of the PBBS or network node can hear you ? If not, then you're probably causing hidden terminal problems. The only solution is to either increase your power or add better antennas so that most people in your local area can hear you when you transmit.

There are many, many, many radios to choose and use for packet radio. The good news is that almost all amateur radios can be used with 1200 baud packet radio. The reason is that the 1200 baud signal fits within the same width as voice transmissions. The problem comes when you want to do something faster than 1200 baud. An important note is that bits per second (bps) and baud rate are not the same. The baud rate can be used to describe the width of the signal. An example of this is any 2400bps or 9600bps telephone modem. Both of these modems actually use 1200 baud over the telephone line, but due to data encoding can fit more bits in every baud. As a general rule, the faster the baud rate, the more bandwidth the signal occupies. The result is that anything faster than 1200 baud typically can't be simply plugged into the microphone jack. These faster baud rates require that you modify the radio in some fashion to get access to the discriminator and transmit element. This is not recommended for beginning packet operators.

As mentioned above, frequency agility is not a high priority. Most packet radio channels are set in an area, and you will eventually find one or two frequencies you use all the time. Why then purchase a $400+ radio that allows you to access every frequency if you are going to dedicate it to one or two frequencies? Something to keep in mind. If you are thinking about operating a network node or service, it is then important to remember that radios that work at your home station are sometimes not adequate for operations in remote or in commercial environments. Many amateur radios cannot handle the commercial RF environments or even handle amateur station environments when the station owner decides to operate several radios on the same band. How radios work and operate is an art and takes understanding to gain the knowledge of how to make it work for you the best way on packet radio. Packet radio requires some different things from normal voice channels, so that is important to keep in mind.

If you are having problems connecting to another station, the following might be a cause: Frequency, Deviation, Volume, or Squelch. Frequency and deviation are sometimes the hardest to check. Find someone with a frequency meter to check your transmit and receive frequencies. A not so accurate method is to get another radio and see what the voice sounds like over that channel. This works for many an FM radio as a check before someone comes over. Deviation is a little harder to check, but there are several inexpensive deviation meters on the commercial market and I have seen some network nodes that measure deviation and report it back in the heard list. More retries are caused in our area by over deviation than any other cause. Over deviation causes your signal to exceed the limit of what the other station is capable of decoding, thus some of the data is lost due to it being too wide, and the entire packet is never exchanged. Each TNC and radio has a slightly different volume setting required. Typically what you think is readable audio coming from your unit is almost always too loud for the TNC, but that depends on the TNC. I suggest plugging a speaker in parallel with the TNC so that you can monitor the signal along with the TNC. When you hear a packet, and it sounds pretty clear, then the TNC should decode it. If it does not - try changing your volume a little up or down. Change the volume setting only after several packets have been heard in this way. If you constantly change the volume, you will not be getting a good guesstimation factor. If your TNC provides true DCD (Data Carrier Detect), then you should be able to operate with your squelch open (lots of background noise). Packet uses DCD to determine when to transmit. Many TNCs don't provide optimum DCD operation, and version 2.0 of the AX.25 Level 2 protocol compounds the problem. If your TNC does not have true DCD, then if you run with the squelch open, the TNC will think that someone is constantly transmitting. During band openings you might even have to adjust your squelch up, in order to eliminate distant signals and noise. You can either upgrade your TNC with a DCD kit or leave your squelch set to a high enough level as to not break until there is actual packet activity on the channel.

The most difficult part of packet radio, I think you will eventually discover, is not the packet (digital) aspect but the RF (radio) side of the equation. Radio technology can be magical at times concerning how it works and if you have someone in your local area that is knowledgeable about radios, be sure to ask their help.

References:

Jones, G. (Ed.) (1993). Packet Radio General Information. Tucson Amateur Packet Radio Corp.

Lucas, Larry, Greg Jones, David Moore. (1992) An Educator's Alternative to Costly Telecommunications. Texas Center for Educational Technology, University of North Texas. Denton, Texas.

Note: *This article was first published in 1984, but is still very pertinent today. The hope of this article is to have outside experts explain the POOP on packet and possibly make it more acceptable.*

Eliminating POOP (Poor Operating on Packet) from Packet

Peter Eaton, WB9FLW and Lyle Johnson, WA7GXD

Overview

A lot of POOP has been discovered on packet frequencies across the nation and around the world! Indeed, in addition to its health and welfare implications, POOP is both unnecessary and oftentimes offensive. While other four-letter acronyms have been used to describe the characteristics of POOP, it is hoped that POOP is sufficiently recognizable by packeteers to eliminate the need to express the others!

What, exactly, is POOP? How does one eliminate it? How can one help others to cause it to not be propagated? The answers to these questions form the basis of this paper.

POOP - What is it?

POOP is simply an acronym for *Poor Operating On Packet*. While it may evoke other thoughts in one's mind, the relationship between those other thoughts and poor operating practices is probably pretty clear and will not be further elaborated upon.

POOP - How does one eliminate it?

In order to eliminate POOP, one must simply not generate it. If it is generated, it will be passed onto packet channels, needlessly clogging them. While there are many varieties of POOP, and it would be impossible to describe them all in this paper, several of the more obnoxious and prevalent forms of it are described.

If you have ever been around a pond, you have undoubtedly heard the loud and constant noise put on by frogs. It seems amazing that so small a creature can make such a disturbance! If you have ever monitored a busy packet channel, you have probably seen plenty of beacon messages. Here again, a large disturbance may be caused.

Beacon features were included in TNC software in the early days of packet when stations were few and far between. Like the frog on the pond, the noises were made to attract attention of like species - in this case, other packet stations. Unlike the frog, who settles down after he gets what he was looking for, many packeteers continue to send beacons, often on crowded channels. Some packeteers contrive clever beacons, to sound bells, clear screens, or print multi-line declarations on the screens of all who can decode the beacon.

<u>**The proper rules governing beacons are simple**</u>:
1) Determine why you need to beacon.
Beacons declaring that you are unavailable, or on vacation, are perfectly useless and mark you as a real POOPer. If the information you are attempting to convey is important, perhaps leaving it as a message addressed to all on the nearest packet bulletin board station (PBBS) is a better alternative. On the other hand, if you are living in tornado alley and see a funnel, an urgent beacon may be appropriate.

(In-search-of POOP) If the purpose of your beacon is to let folks know you are around and want to connect, it may be better to just turn on the radio and let your TNC decode a few packets from other stations. This way you can see who is on and then simply send a connect request rather than a beacon.

Many new TNC software releases include an MHEARD function, allowing you to see the contents of a buffer containing the last several packet stations heard by your station.

If you are convinced that you must transmit without listening for a few minutes (or if the channel really does appear dead), dropping into UNPROTO mode (CONVERSE mode from COMMAND mode without first connecting) and typing a short CQ message (which may be as simple as a carriage return if UNPROTO is set to CQ) is preferable to beaconing one.

2) Compose the briefest possible beacon text.
Cute beacons that fill a screen, sound bells, or clear screens will only mark your station as obnoxious. It is a classic way to lose friends and increase your count of enemies!

3) Use BEACON AFTER rather than BEACON EVERY.
If the channel is busy, one-way broadcasts (which is, after all, what a beacon really is) are not welcome. It's bad enough to try and maintain a connection through a digipeater or two without having a channel clogged by transmissions from unattended stations that come on the air every few minutes. Beacon AFTER with a value of thirty minutes will assure that you do not add to busy channel bedlam.

4) Don't send beacons more often than every thirty minutes, preferably less frequently.

5) Digipeat beacons with care!
Digipeating may cause a large number of local packeteers to be subjected to screens full of your beacon text. This may be desirable. Then again, it may not. Consider your motive and objective for your particular beacon, then set up the path.

Bull POOP

Try entering a field containing a bull. While many bulls are mild mannered, some are very territorial and will chase you away. The same is true of a packet BULLetin board station. Many are mild mannered, aware of other packet stations on the channel and content to share the channel with them. Others, however, are not. They will chase you away unless you came to feed them. They do it quite simply, and often are ignorant of their ferocity.

A skilled matador, however, can soon tame a ferocious bull. So can the operator of a PBBS tame his BULL. The keys are TNC setup files. Most PBBS software contains a file (or files) describing the characteristics of the TNC(s) attached to the computer serial port(s). The magic commands are PACLEN, MAXFRAME and DWAIT.

If a PBBS is operating on HF, PACLEN should be fairly short, perhaps 40 or so. Since this parameter describes the length of the information field, not the header and control bytes, a setting in excess of 80 (the length of one line on most computer displays) is probably the longest needed.

MAXFRAME can be the cause of a lot of useful bandwidth reduction. If the PBBS is on a channel shared by other users, MAXFRAME 1 is reasonable. We have heard PBBS's sending packets of many frames to stations that were having a hard time decoding anything, and the channel was reduced to uselessness for other stations. Similarly, we have often heard PBBS's on HF sending long packets of multiple long frames, getting an ACK on the first one only (if any), and repeating the process over and over. Computers are infinitely patient, but humans wanting to use the channel may not be!

DWAIT is perhaps the strongest medicine to apply to an overly possessive BULL. PBBS stations should set DWAIT to 320 milliseconds. For a TNC 2 it is DWAIT 32.

If you are not the owner of a BULL, but venture into territory where one lives, you can help tame the beast! The following suggestions are highly recommended:
1) DO NOT DX A PBBS! In this case, DX means multi-hop digipeating or networking to a PBBS on VHF.

2) Don't send the PBBS a command before it has responded to your previous command! Hitting a key twice (or hitting it harder!) ***WILL NOT*** improve your chances of getting through! The nature of a packet system is that the message gets through accurately, or not at all. Sometimes it may take a while, especially if a digipeater or HF link is involved, but it will get through. If not, you will get a ***DISCONNECTED message.

Untimely POOP

POOP can't easily be made timely, but TNCs can! And TNCs that aren't timely can sure contribute to the level of POOP on a packet channel!

In the January, 1986 issue of PSR Quarterly, Tom Clark, W3IWI, made a convincing argument for the setting of the DWAIT parameter for all packet operations.

User Type	**Time mSec**	**TNC1 DWAIT**	**TNC2 DWAIT**
Digipeaters	0	0	0
Keyboard Users	160	4	16
PBBS, Hosts	320	8	32
File Transfer	480	12	48

Digipeaters wind up with the highest priority. Since these stations are the most susceptible to collisions, and generate the most congestion on a retry, they deserve first shot at an empty time slot.

Keyboard users, operating in a keyboard-to-keyboard QSO, generate little traffic. After all, one can only type so fast! They get the next priority. PBBS and host stations generally produce a fair amount of data out for a little data in. Thus, the keyboarder has priority getting into the PBBS, but the PBBS waits for other keyboard users before dumping what will probably be a longer packet onto the channel. File transfers, generating the most data and hence requiring the most bandwidth, are requested to be more polite and give other users a fair shot at the shared channel. Thus, they are held off the longest.

Wide adoption of this scheme may not significantly reduce congestion on a channel, but it should help the channel operate on a fairer basis than otherwise.

Snake POOP

A snake has a fairly unique characteristic. A snake has no ears! Too many packeteers seem to have the impression that, by connecting a TNC to the speaker jack on their radio, they don't have to have a speaker connected! The results can often be observed. Excessive retries on a channel because the antenna isn't oriented properly leads to multipath and poor reception. The other end of the link simply "goes away" for no apparent reason (unless you are listening!). The other station is over-deviating, or another user on another mode, or... And, on a shared-mode channel (or shared repeater), packet can get a bad name in a hurry!

Kangaroo POOP

A kangaroo jumps around. If you have long files to transfer, you should jump around, too! A busy channel during the early evening hours is not the place for file transfers, automatic message forwarding or similar bandwidth-hungry procedures. What can you do? Jump off to another frequency, perhaps. If this is not feasible, set your alarm for 3 AM and jump to another time, eating up the channel then.

POOP - the final scoop

The ultimate means to eliminate POOP is to SCOOP! By means of the SCOOP, no one will ever be able to detect packet POOP emanating from your station! SCOOP means Setting COrrect Operating Parameters. If you heed the advice to avoid POOP given above, this final measure will permit you to have a full clean-air rating! — Happy Packeting!

A Simple Guide to Better Packet Operations

TNC Settings : This list is the TAPR recommendation for TNC settings when using the current 1.1.8a TNC firmware on VHF. If you would like information on how to get a copy of 1.1.8a, contact Tucson Amateur Packet Radio at (817) 383-0000. If you are not using 1.1.8, set all settings that are implemented on your TNC. 1.1.7 implements a number of new features which dramatically improve channel operation.

ACKPRI	ON	ACKTIME	14
DEADTIME	33	DWAIT	33
FRAC	8	RESPTIME	0
SLOT	3	TXD	33

PACLEN 64-255 depending on channel quality.
MAXF 1-7 depending on channel quality.

People who get the most discouraged when operating packet have problems with either the way their TNC is configured or are trying to make packet work over a poor path to begin with (Handi-Talki in an apartment). If you are having problems, ask a packet operator who knows the parameters for your TNC and see if something is incorrectly set. Sources for problems are : TNC settings, better signal needed, radio & TNC deviation problems. Any of these or a lot more might be your problem.

Common Courtesy - Remember that packet radio is a shared resource. Typically whatever you do when operating packet affects someone else or vice-versa. Operating points to keep in mind are: Beacon Usage, Digipeating, DX BBSing, and Network Utilization. None of these items are set in stone. Each situation requires that you use good judgment on how you operate. Packet can be a lot of fun or it can be a nightmare, all determined by channel operations.

Local traffic - Packet can be thought of much like a voice repeater. If you have a local connect to make, don't stay on a congested packet channel to do it. Move to a less crowded, usually non-network or BBS frequency to conduct local exchanges.

Beacons - Beacon usage can be used in good and bad ways. Typically, if you HAVE to beacon, set your beacon for longer than 10 minutes. Remember that your beacon uses the UNPROTO setting to determine its path. Unless you have a VERY good reason to BEACON through DIGIPEATERS, don't do it.

Digipeating - Digipeaters were originally implemented to take up the slack for nonexistent networks at the time AX.25 was originally defined. Digipeating has been proven to be one of the worst methods of packet connections, but when it is the only one available it is better than nothing. When you can use a network over a digipeater, do so.

Network Utilization and DX BBSing - The thing to remember when using networks is that what you do at the far end now affects operators that you can no longer see, who are many network hops away. A good rule of thumb about using networks is to ask someone who already uses the network about what you are doing, or experiment when you know channel access is low, so that you will not affect very many people.

DX BBSing is the most seen example of this problem, since BBSs are well established, stay on the same frequency, and operate 24 hours a day. What is on your LOCAL BBS will be the same on a distant BBS, with the exception it might take a day longer to get to your system. Mailbox/BBS systems are designed to be used locally by users and pass messages and traffic during low channel usage times. Allowing Mailbox/BBS systems to do their jobs, by not circumventing them using a network, allows better utilization of the packet channels.

The Elmer Solution - Amateur radio operators have had ELMERs around for all times, it seems. A packet Elmer is one of the best ways to get on packet with the least amount of problem and pain. Packet Elmers are people who have been on packet and know how to get around the local packet frequencies and networks. The easy part of packet radio is connecting the TNC to your radio. The hard part is actually using packet to accomplish something besides seeing screen after screen of stuff you don't understand. **An Elmer can show you some very quick and productive ways to start enjoying packet right away.**

Beginners' Frame: Command Settings

Jim Neely - WA5LHS © 1991

Packet radio is a fascinating mode of communications and holds much interest and not a little puzzlement for newcomers. We will recommend a few more places to gain some additional data and hopefully a little enjoyable and informative reading.

Packet communications is based on the AX.25 protocol. If you'd like to dig into the bits and pieces of this protocol and are comfortable with computer terminology, ARRL publishes a book called "AX.25 Amateur Packet-Radio Link-Layer Protocol." In all fairness, you don't have to know all the bits and pieces of the protocol to use and enjoy packet communications. Another book that is quite helpful and will provide more of the background and history of packet radio is "Your Packet Companion," by Steve Ford, WB8IMY. This book is published by the ARRL and should be available in the local ham retail stores or directly from the ARRL. Yet another ARRL publication is "Your Gateway to Packet Radio," by Stan Horzepa, WA1LOU.

TNC (terminal Node Controllers) use a set of commands to tell them how they are to operate. These commands are one of the areas of great confusion to beginners who have no basis for knowing the settings for these commands. Figure 1 and 2 are lists of commands that are used in my TNC2 and PK-232 TNCs. If a particular command isn't listed, the factory default is used.

Just because I used a particular command doesn't mean that is the only way to do it, or even the correct way. This is a set of parameters that I have worked out. The reason for these suggestions is just as a starting point. After you make things work, you can change one parameter at a time and see the effect and determine for yourself if you want to use it the new way or the old way.

One recommendation I do have is that you turn all beacons OFF. Beacons served a purpose when there were very few people on packet and you wanted your friends to know when your station was on. Today, we have trashed the 145.01 frequency very badly with too many stations, too many digipeaters and NetRom/TheNet stations using the user frequency for network communications. There is plenty of QRM on our packet frequencies without new users adding useless beacons to the confusion.

We'll add a few more tidbits to help in operating and learning to use packet communications in the next issue.

Figure 1 - TNC2 Set Up

AWLEN	8	PAR	0
AX25	ON	CBELL	ON
CMSG	ON	SERS	0
CONSTAMP	ON	DAYSTAMP	ON
ECHO	OFF	HEADERLN	ON
MON	ON	MDIGI	ON
MSTAMP	ON	MYCALL	WA5LHS
CT	de WA5LHS, Jim, Austin		

Figure 2 - PK 232 Set Up

8BITCONV	ON	MAXFRAME	4
ALFDISP	ON	MBX1	
ALFPACK	OFF	MBX2	
AX25L2V2	ON	MCON	0
AXDELAY	0	MDIGI	ON
AXHANG	0	MFIL	00000007
BEACON	EV	MONITOR	2
BEACINT	0	MRPT	ON
CASEDISP	0	MSTAMP	ON
CHECK	30	MYA	JIM
CHLABEL	OFF	MYC	WA5LHS
CMSG	ON	PACLEN	80
CONALARM	OFF	PACTIME	AF
CONSTAMP	ON	PACTIMINT	10
CPACTIME	OFF	PASSALL	OFF
DAYSTAMP	ON	RELINK	OFF
DWAIT	16	RESPTIME	10
FRACK	3	RETRY	7
FULLDUP	OFF	SQUELCH	NEG
HBAUD	1200	TXDELAY	75
HEADERLN	ON	TXECHO	OFF
HID	ON	USERS	10
ILFPACK	OFF	VHF	ON

Beginners' Frame: Basic Problems

Jim Neely - WA5LHS © 1991

In the last Beginner's Frame, I covered a few basics of packet, various TNC parameters for the TNC2 and the PK232, and hinted at a few other areas to be covered in later columns. This time, I will cover some of the problems that beginner packet operators face when they start working packet.

One of the things I mentioned in the last column was the problem in packet radio termed 'hidden transmitter' or 'hidden terminal'. One example of this problem can arise from using a low power transmitter, such as a handy talkie. Your station may be able to hear and hit a node with your HT, but another station on the other side of town might not be able to hear your signal and his TNC will transmit when yours does, causing a collision with both packet transmissions. In this case, neither one of you will get your packet through. It is a frustrating experience for everyone involved. Typically 'hidden terminal' problems arise when two stations can hear the central station, but cannot hear one another. By the same token, you don't need to run a great deal of power. Remember the FCC regulations call for the use of no more power than is necessary.

One of the nice things about the packet protocol is that if your receiver is hearing another station transmit, the TNC will not allow your station to transmit and cause a collision. Built into the TNC's is a somewhat random delay between transmissions, which you can control by using the DWAIT parameter. If you find yourself having collisions with the digipeater you are using, perhaps a change to the DWAIT parameter will help.

I talked to a local club in Austin about packet recently and some of the questions and comments concerning packet that came up were interesting. Probably the content, or lack of content, in the TNC manuals has a lot to do with the tone of most of the comments.

One comment that was made concerned how you disconnect from someone when you have finished a QSO. Most TNC2's require a "Ctrl-C", that's a CONTROL and C pressed at the same time. The

Ctrl-C should produce a "CMD:" prompt on the screen. After you have the "CMD:" prompt, you can then give the TNC any command you'd like. The shorthand for disconnect is "D". Remember you have to press the "Enter" key for the TNC to see your command.

Another area that raised questions was about aliases for digipeaters. All digi's have a call, usually with an SSID other than -0. They usually also have an alias. In the early days of packet radio, the aliases were the same as the airport identifier. In Austin, the club digipeater alias is AUS. This same pattern is followed in many areas. It is also common for the last two digits of the frequency to be appended to the alias. The Bastrop digipeater on 145.09 alias is BAS09. Please keep in mind that digipeaters are not networks and shouldn't be used as such.

There were several questions about how you find the path from your station to a distant location. One method of finding a path is to watch the screen to find someone else getting into the desired location and use the data on the header line to find some or all of the path. The other is to contact a local packet operator and ask about operating procedures and network connections. An ELMER for packet can really help.

Beginners' Frame: Packet Software

Jim Neely - WA5LHS © 1991

One of the often asked questions by beginners is about software for packet. There are a number of communication systems available that will work quite well. Many people are great believers in using the same software for packet that they use for modem communications. One of the problems with this theory is that you are basically using sophisticated software in dumb terminal mode.

In the last few years there have been several systems written specifically for packet. Some of these work better than others and some of them suit your particular operating habits better than others.

There are several systems designed to take advantage of specific TNC's and their peculiarities. One of the most popular of the multi-mode TNC's is the AEA PK-232. This was also the first of the breed and is widely used.

My PK-232 is serial number 1278 and has had the firmware, that is, the program chips, upgraded several times. The PK-232 was the first TNC to make use of the host mode. AEA marketed a piece of software to take advantage of the PK-232 features called PC PAKRATT. This is one of my favorite packet programs and is the standard by which I judge other software even if it is getting long of tooth.. There are several things I'd like to see different in PC PAKRATT, but overall it is a very well thought out piece of software. One of the better features of PC PAKRATT is a screen with all the parameters on screen at once. This makes it quite easy to see the relationship of certain parameters and makes experimentation much easier.

Necessarily I am limiting my discussion of packet software to the PC compatibles. Perhaps, in the not too distant future I can try out some of the Macintosh software.

In the last few weeks, I heard about a new packet system called PACKET-GOLD Ver 1.1. This is marketed by Interflex Systems Design Corporation, P.O. Box 6418, Laguna Niguel, CA 92607. PACKET-GOLD was written by Lynn Taylor, WB6UUT, and Jeff Towle, WA4EGT. At

$59.95 it is well worth the price and has a number of nice features including a well printed 72 page manual. They also have an upgrade policy for PC PAKRATT users. PAKRATT users can send in the original disk with the AEA label and $39.95 for a copy.

PACKET-GOLD starts out by configuring the TNC to suit your preferences from a file of parameters. It figures out what configuration the PK-232 is in and changes it without restarting the TNC, therefore saving any messages in the TNC buffer. This is a feature all TNC software should have.

One of my pet peeves with packet software is what they show you on the screen and how it is presented. PACKET-GOLD is reasonably configurable to your requirements.

One of its whiz bang features, which I am not as thrilled about as the authors, is the ability to watch what else is going on while you are connected. This is shown in a configurable window at the top of the screen. You can configure the size of this window to suit yourself. The recommended size is 6 lines on a 25 line screen and 10 - 12 lines on a 43 line screen. Frankly, I'd rather have the screen space for the received data, so I configured for zero lines. You can also use the zoom command(alt- z). The program uses about 4 - 5 lines for some nice-to-know information which shows up between the network view and the received data. This data shows how many pending packets are outstanding and how many retries have been made. It also shows the status of the TNC: Connected, Waiting Acknowledgement, etc. There is only one line for the transmit data, but the line is editable, which is nice. Another of the nicer features is the ability to capture text from the screen and store it in a file with your choice of name. One of the things I do with this type of feature is capture portions of messages and pass them along to others on different networks. The only complaint I had was that I didn't figure out quickly how to choose files in the other directories. Unlike some of the other software which requires that you enter the name of the file to be transmitted in an area that links it to a function key or that you name the file to fit the program's requirements, PACKET- GOLD shows you a window with the file names. You then pick the file with the cursor keys and press enter to send it.

There are a number of other features which I either haven't had time to play with or had no interest in. One of those that fall in the latter category is the ability to check your mailbox, if you have the firmware upgrade that allows that. I'm one of those people that thinks we need less, not more, mailboxes on our busy frequencies.

PACKET-GOLD supports any communications port, handles Siam, Navtex, CW, Baudot and Amtor. You can also do a file transfer while having another QSO, or shell to DOS for other programs. One of the other features of PACKET-GOLD that deserve mention is the ability to go to a previous QSO with the touch of a function key. PACKET-GOLD has multiple connect capabilities if you are into talking keyboard to keyboard with multiple stations. Also, planned for a future upgrade is a conference mode. PACKET-GOLD also is completely configurable for screen colors and 25 , 43 or 50 line screens with the correct video capabilities.

Another packet system that is nice and has some special features is LAN-LINK by Joe Kasser, W3/G3ZSZ. LAN-LINK has built in capabilities to chase and connect to SAREX robots. This might be of interest in the upcoming SAREX flights. LAN-LINK is configurable to TNC1, TNC2, MFJ1278, PK-232, KAM and KPC TNC's. It is available from Joe at 11421 Fairoak Drive, Silver Spring, MD 20902, for $35.00 or is available as shareware on some landline BBS's. It is also part of the TAPR software library.

Beginners' Frame: STS/MIR Packet

Jim Neely - WA5LHS © 1991

With the many the shuttle flights using radio, there has been fresh interest in working satellites using packet. The shuttle flights are fun and interesting, but short lived. The Russian spacecraft MIR is currently running a packet personal message system most of the time.

I haven't seen any publicity regarding coming SAREX payloads on coming flights, but there will surely be more flights with hams on board.

There was some confusion about the SAREX stations and how to work them. The STS-35 flight used 145.55 as a primary transmitting frequency and they listened on 144.95. There are at least two things to notice about this combination. The first thing is that the frequencies are 600khz apart, so the minus offset on your transceiver should be used. The second thing to notice is that the frequency is the same as one of the popular DX Cluster frequencies. I heard a few complaints about the use of this frequency. Most of the complaints were about people calling the shuttle while it was not in view, or just ragchewing on the frequency.

The DX Clusters moved to a little used frequency so they would not interfere with the BBS and keyboard users. The least we could do is return the favor and not clutter up a frequency ragchewing on top of a DX Cluster. Most of the cluster users were also trying to call the shuttle, so they had little gripe about the use of the frequency by the shuttle.

The complaint about calling when the shuttle was not in view is an easy one to solve. AMSAT has two excellent tracking programs. Quiktrak by Bob McGwier and Instant Track by Franklin Antonio are excellent tracking programs. Instant Track is faster but Quiktrak has printing capabilities. These programs are inexpensive and should be used to determine when the shuttle is in view. It is quite simple to get the input data, called Keplerian elements, from your local BBS, CompuServe or Dallas Remote Imaging Group BBS. There is usually a set of preliminary data available prior to the launches which should at least get you close.

Screen from InstantTrack

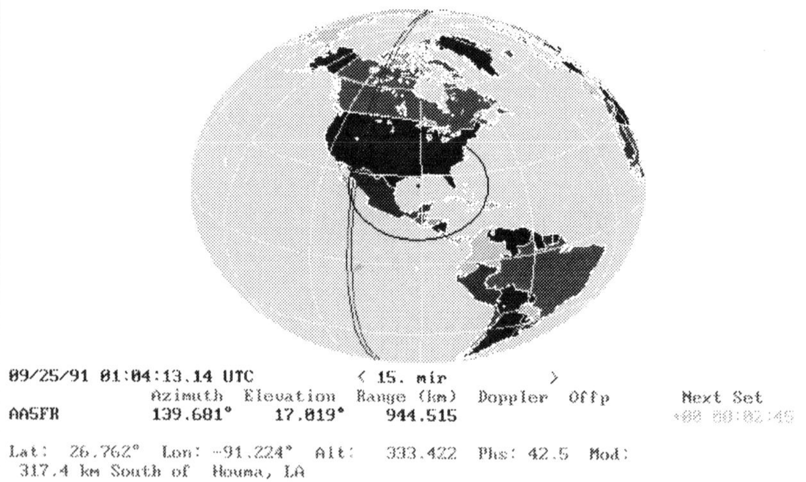

One of the other ways to handle the problem of calling when there is no one to hear you, is to use Joe Kasser's Lan-Link. Lan-Link has a SAREX mode, in which it will wait until it hears the call you have entered, then it will do an automatic connect, get the QSO number and log off. I think it is pretty neat to set it up to have my machine call his machine, since the passes over Central Texas were during the work day, or while I should be sleeping. I have been known to get up at strange hours to chase a satellite, but if the computer can do it, probably better than I, why should I interfere.

The MIR is using the call U2MIR-1 for the personal message system and they are operating on 145.55. The MIR operates in the usual simplex modem. The cosmonauts aboard the MIR are Musa and Victor, if you'd like to try to leave them a note on the message system. (in 1991) I have seen him connect to a BBS and have captured conversations he has had.

Remember that these low earth orbit (LEO) satellites are only within hearing range for 15 - 20 minutes per pass, depending upon the height above your horizon. I saw someone complaining on FIDONET about the fact that the inclination to the equator was so far south they only had a 15 minute window to hear it. That's about the best you can do.

Phase 3D Satellite (scheduled for 1995)

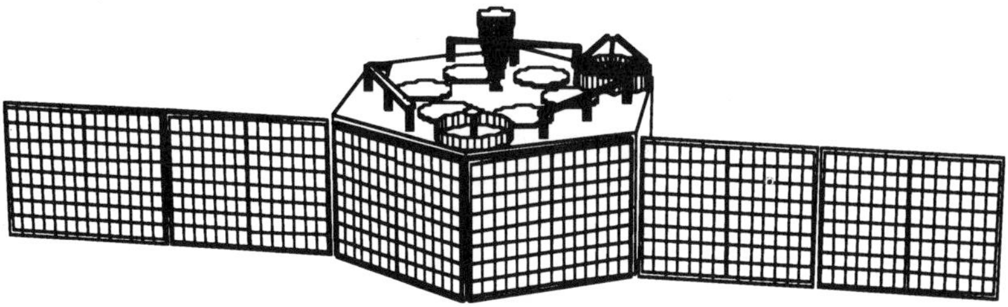

Working the MICROSAT is a little different and takes some additional equipment and a primer on that aspect of packet is beyond the scope of this column.

The DOVE satellite, Oscar 17, does transmit its telemetry on 145.845 using the usual 1200 baud packet protocol. Joe Kasser has a program called What's-Up that will read and interpret the telemetry data on the screen.

If you haven't tried working a satellite, try the MIR or listen for the DOVE. I quite often leave the radio on either the MIR or DOVE frequency when I'm not there to watch it. I have quite a bit of DOVE telemetry in files to play with at my convenience.

The addresses for the software mentioned are:
Joe Kasser W3/G3ZSZ, 11421 Fairoak Drive, Silver Spring, MD 20902

AMSAT/NA
850 Sligo Avenue, Suite 600, Silver Spring, MD 20910-4703
(301) 589-6062 (9AM - 5PM Eastern).

Have fun on the birds. I hope to see you there sometime soon. 73.

Beginners' Frame: Packet BBS

Jim Neely - WA5LHS © 1991

Anyone with a TNC on the air in most areas of the US and many foreign countries has seen the Packet Bulletin Board System (PBBS) in action. These have been confusing critters to some old timers much less newcomers to packet.

A PBBS can be recognized by it's beacon which often has a line with "MAIL: WA5LHS WB5DDP N5LJF ALL". Of course the calls can be any call.

These PBBS stations move messages around the country with surprising efficiency. I say surprising because it is always a surprise as to how fast or how slow messages can move. It is not unheard of to move a message from coast to coast in a day or two, but by the same token, it may take much longer. The messages are moved by a combination of VHF/UHF and/or HF relays.

All the preceding is background information so you will have a better idea what the PBBSs are all about. Connecting to a PBBS is no different than connecting to any other station. You connect to the station call sign. Some of the PBBS software wants to know something about you such as your name, home BBS, zip code and several other possibilities. This is to make them appear more user friendly in some cases and to help route messages back to you in other cases.

Most PBBS software has a help system which is accessed by the "?" or "H" commands. Some of them offer command-specific help, too. This is usually explained in the help message you get from the system. Probably the most important thing a PBBS user needs to know is that you can't hurt anything by something you do on the air. The PBBS software in use today goes through almost constant upgrading to improve its facilities, make it easier to use and also make it more bullet proof. The nearly universal commands on PBBSs are R for Read a message, L for list messages, S for send a message, K for kill a message and B for Bye.

The "L" commands lists the messages since your last sign on in most cases. There is also a "LL" command. It is followed by a space and a number. "LL 10" means to list the last 10 messages headers. These headers tell you who they are to and from and the subject matter. Some headers have the size of the message and date and time they entered the system.

You may read a specific message by using the "R 2560" command where 2560 is the message number you want to read. Some software will allow you to enter more than one message number on a line such as "R 2560 2600 2605 2606".

To send a message you use the "S" command which may be modified by adding a second character to the "S". You use "SP" to send a personal message which can only be read by the addressee. You can also use "SB" to send a bulletin. In all cases the "S" command is followed by a call or description such as :"S N0CCW", "SP W9DDD", "SB ALL @ N5LJF". The "@" symbol means "at". You can send to "ALL @ ALLUS" but this is discouraged unless it is really appropriate. It is rare that someone in Maine is going to buy a TNC from someone in California.

After you send the "S" command the PBBS will ask you for the subject of the message. Enter a short description of the message.

The system will then ask for the message and often tell you to enter "/EX" or "CTL Z" to end the message. You should probably enter the "/EX" or "CTL Z" on a separate line. The "CTL Z" is done by pressing the "CTRL" and "Z" keys at the same time. You exit the PBBS by using the "B" command. If you should forget to exit, most software will disconnect you after a few minutes of no activity.

This short treatise on the PBBS systems should only whet your appetite.

Beginners' Frame: TNC Alternatives

Jim Neely - WA5LHS © 1992

The magazines have had articles on a couple of TNC alternatives in the last few months. 73 magazine wrote about PMP otherwise known as Poor Man's Packet and QST had an article on Baycom.

These are both kits which are modem-only hardware devices that use the computer to do the work usually done by a TNC. Both kits are pretty straight forward to build and work pretty well.

The Baycom kit works using the serial port and the PMP uses the parallel port. PMP can draw it's supply voltage from the serial port, too, so a battery or other power supply is not necessary. The Baycom software was written in Germany and the documentation was translated to English. There is pretty good help built into the Baycom software, which consists of 2 programs. The first is a terminate and stay resident program that does the job a TNC normally does. Baycom itself is the terminal program that the user sees. It is reasonably user friendly and easy to use.

The first time I printed out the documentation, I didn't get it all and had a problem trying to figure out how to capture or transmit a file. The commands for these are not absolutely intuitive until you read the documentation, then it makes quite good sense. The command to send a file is Read the file and the command to capture a file is Write the file. These are pretty obvious once you find them and see how they work.

I would suggest that anyone who wants to play with Baycom read the documentation all the way through. Some of the translation from German to English is little quaint, but quite readable and very well done. Also, as in the case of capture or sending a file, the terminology may not be obvious until you read the documentation.

The August 1991 issue of 73 featured the Poor Man's Packet as it's lead article. This article is the kit building documentation and is quite well written. Kevin Feeney, WB2EMS, and Andy Payne, N8KEI, designed the hardware and wrote the software as a necessity, since a TNC was out of a student's budget.

The PMP is probably a little simpler kit to build with fewer parts and a little less crowded board. The PMP software is pretty easy to use, also. You need to edit the configuration file and get your call in place and the paths that you use, but that is well explained in the documentation. Alt-H brings up a single-screen help text that explains all the commands. This is quite straightforward. One thing to remember in the software documentation is that the discussion of the 1200 baud modem refers to the PMP modem, unless you have ideas of designing and building your own. Both Baycom and PMP are well suited for portable operation and have been used this way quite successfully. Another consideration for both kits is that they only work with IBM compatible computers. I did see a message in the packet bulletins that neither worked with an NEC multispeed laptop, but this is the only problem I have seen. Both modems are fun to use and easy projects to build.

A complete kit for PMP is available from the authors for about $45.00. Don't expect immediate delivery as they are doing this as a favor, because the chip they use is not available everywhere and I think they may have been surprised by the number of orders they received.

The Baycom kit is available as a bare board, board with parts or completely assembled and tested. Prices range up to $89.00 for the assembled and tested kits.

PMP Kit and disk orders should be directed to Kevin's address. Kevin Feeney, WB2EMS, 468 Hines Road, Newfield, NY 14867 Please don't telephone. We aren't prepared to handle phone call regarding PMP. For the printed circuit boards and the miscellaneous parts kits, please specify the laptop version or the home station version. Please make your check or money order in US funds drawable on a US bank. No assembly instructions are included: refer to the 73 article.

The address for A & A Engineering, who handles Baycom is: 2121 W. LaPalma, Unit K, Anaheim CA, 92801, Phone (714) 952-2114.

NOTE: Since this article was written, Tigertronics offers an assembled unit for around $50.

Beginners' Frame: Packet and HTs

Jim Neely - WA5LHS © 1993

One of the most helpful people I have come across in packet radio is Norm Sternberg, W2JUP. Norm wrote the original PK-232 manual and for several years he was the AEA support person on Compuserve. There have been many discussions about using handheld radios on packet and when I saw the following piece on Compuserve, I asked Norm for permission to use it for this column. Norm graciously granted me permission to quote him and I'd like to thank him in public for his generosity.

[This piece was written about nine months ago in answer to some questions from packet newcomers in the Long Island area. I think it still applies to any area in which there's lots of packet radio activity. - Norm, W2JUP]

Now that I've read at least a dozen PBBS bulletins on the subject of using handy-talkies on packet radio, I feel like chiming in with my own ideas on this subject.

I grant that some low-power "HT" packeteers can probably work their local PBBS, and probably work more-distant PBBSs through various nodes and switched. But, in my not-so-humble opinion, using an HT for packet radio, at the average ham's location, is equivalent to redefining the meaning of "HT".

In too many cases, "HT" can mean "Hidden Terminal"!!

The Commission's Rules (Part 97.313(a) for the newcomers) state hat "An amateur station must use the minimum transmitter power necessary to carry out the desired communications". This means that the packet station must use enough power to establish and maintain communications, especially when working a packet BBS station. The key word here is "maintain", as in keeping a successful packet link or connection.

For the benefit of newcomers, a "Hidden Terminal" is a packet radio station that is NOT heard by all of the stations on a given radio channel. You may become the "Hidden Terminal" whenever OTHER packeteers on the same radio channel CANNOT hear your frames.

Whenever you send a packet, any station that "hears" your packet frames should have his/her TNC's DCD respond to your packet. The DCD signal inside the normal packet controller PREVENTS the TNC from keying the radio and sending packet frames UNTIL the channel is clear. This corresponds to the commercial practice known as "CSMA" or Carrier Sense Multiple Access.

Without the benefit of your signal's resultant DCD holding the other station's packets off, the other station transmits directly into YOUR packets.

Whenever this occurs, the PBBS or node or switch hears BOTH packets at the same time. The PBBS or node or switch cannot decode either station's packets. When this happens, both your station and the other stations send the same packets over again based on your "retry" counters. Each time you DON'T get an "RR" or at least an "RJ" frame from the local PBBS or node or switch, your TNC will retransmit the same packet frames over again. Each time your TNC retransmits and the other stations on the channel CAN'T HEAR YOU, they also retransmit!

The result? Collision City! Stations time out or disconnect on retries and the channel turns to a colossal bag of worms. I wonder how many of these "low-power" packeteers have ever thought about the fact that their HTs may really be creating "Hidden Terminals"?

Let's assume that you CAN use your HT to log on to your local PBBS or node or switch. Let's also assume that your target system is running some multi-user software such as the ubiquitous MSYS or FBB or DXCLUSTER programs. Let's assume that there are numerous other packeteers on the same radio channel, all connected to the same PBBS or Cluster or node or switching station. Now here's my question: how many of the other packeteers within that PBBS or Cluster or node or switch's operating radius on that same radio channel can HEAR YOU each time your TNC sends frames???

Think about this for a moment...and then decide if your HT may possibly be creating the notorious "Hidden Terminal" syndrome!

During endless hours of monitoring the packet channels, time after time I see numerous packet stations sending frames and not getting the desired responses from the distant station. In most cases, a small amount of knowledge of these station's locations shows that the problem may be the "Hidden Terminal" syndrome.

There's only ONE cure for this "Hidden Terminal" problem: increase your power to the point where the MAXIMUM number of stations on the same radio channel will HEAR YOU!!

In an ideal packet radio world, we would all be using full-duplex repeaters to handle our packet frames to all other stations. The full-duplex repeater would go a long way towards guaranteeing that ALL stations on the channel HEAR ALL packet frames from ALL stations!

The real question is **Do you have enough signal on the radio to ensure that all stations within range of the PBBS or Node or Switch can hear you ?** If not, then you're just one more problem-maker on the channel.

Those of you who choose to run HTs on packet radio need to evaluate your positions, consider your location and your antenna, and try to understand this "Hidden Terminal" problem.

If you insist on running low power in this complex packet radio environment, then you may really be part of the problem when you should be part of the solution.

Norn W2JUP

HF TNC Settings

Besides the obvious fact that HF and VHF packet frequencies are different, certain settings on your TNC must be changed for you to have effective communications. The commands shown below are for the AEA PK-232. Other TNCs have slightly different commands:

VHF OFF Changes the shift of packet tones to 200 Hz (HF) or 1000 Hz (VHF).

HB 300 Data transfer speed must be 300 baud below 28 Mhz. Although you can operate 1200 baud HF on 10 meters, the de facto standard is currently 300 baud. Standard TNCs are set by dip switches on the back of the unit.

PACLEN 20-60 These are general ranges that are dependent on the quality of the path and level of QRM/QRN/QSB. Obviously, these are more of a problem on HF than VHF. Unless you have a great link and a nearly empty frequency, PACLEN settings greater than about 40 can be very frustrating.

DWAIT 0 On VHF this delay is not only gentlemanly (don't grab the frequency as soon as you can every time) but it also allows efficient operating. HF packet is such a jungle that you may never transmit if DWAIT is much above 0 !

MAXF 1 As a rule, the better the path and the lower the retry count, the higher the value of MAX that makes sense. If retries are high, your throughput can be improved by reducing MAXFRAME. Seldom will be the case when MAXF should be higher than one when operating HF.

If your TNC has a Data Carrier Detect (DCD) threshold adjustment, it is critical that this be set to the "flicker" setting, i.e., the DCD light flickers on and off. If your DCD setting is adjusted too low (DCD not lit), your TNC will think there is a clear frequency and transmit even when there is other packet activity on the channel. This is especially complicated on HF by the fact that what appears to be a weak signal to you can be a very strong signal to the other end of your connect. The result is that the probability of collision is greatly increased and the chances for effective, efficient communications are reduced. If your DCD is set too high (DCD always lit), then you will never transmit; the connected station won't see any ACKs from you and he will think you have disappeared into the ether.

Watching the computer screen on 20 meter packet, we sometimes see folks complaining that they don't like HF packet because of the flaky paths and the retries. More often than not, the packets with the complaints have PACL at 128 and MAX = 7 (and probably with the DCD misadjusted)! With the right TNC settings, HF packet can be highly enjoyable and can offer packetters the opportunity for nationwide, and even worldwide, vistas.

Another important operating tip is after you establish a connect, don't stay on the calling frequency. QSY off to an unused frequency, which you might find has a lot fewer retires.

A Tutorial for Using the Packet Channel on APLINK

Jim Jennings, KE5HE

APLINK, a mailbox program written and maintained by Vic Poor, W5SMM, is a system which provides for an A(mtor) P(acket) LINK. Normally, AMTOR is used on the HF side of the system and PACKET is used on the VHF/UHF side. A user connected to either side has access to the entire message base, therefore a user may leave a message on AMTOR which will be picked up by a user on PACKET.

There is a loosely formed network of APLINK stations over the world, primarily in the U. S., who receive and forward traffic. Much of the traffic is formal NTS traffic. However, there is a significant amount of private traffic handled, especially to some foreign countries. You are probably aware of the APLINK traffic associated with Desert Storm.

The principal APLINK stations that I am aware of, at least as far as traffic is concerned are: W1FYR.NH, WA1URA.IN, K4CJX.TN, KP4GE.PR, TG9VT.GTM, KE5HE.TX, N0IA.NV, W7DCR.OR, AND WA8DRZ.CA. All forwarding on AMTOR is initiated manually. All of the above stations scan several of the HF bands, some scan 80 through 10 meters including the WARC bands. Each APLINK MBO (mailbox) may forward automatically to one PACKET BBS from its PACKET channel. Any number of PACKET BBS's may poll or initiate forwarding with an APLINK MBO.

Upon connecting to an APLINK MBO through the PACKET channel, you will find that the system responds to a limited subset of the commands normally available on a BBS. There are a few differences, the main one is that there is no KILL command on APLINK. You may CANCEL a message if it was originated by you. Once you read a message, it is marked as having been read. It remains in the system for 24 hours, at which time it is archived.

The following table is a list of the APLINK PACKET Channel Commands:

B Logoff the BBS.
Just disconnecting does the same thing.

H Sends Help file.

I Sends the Information file.
L Lists all messages.
L # Lists all files from message #.
LB List all bulletins.
L< call Lists all messages to <call>.
If <call> is omitted it lists all messages from you.
L> call Lists all messages to <call>.
If <call> is omitted it lists all messages to you.
LH List all help files.
LR List all users who have signed into the system in the last 24 hours.
LOGIN call Logs you into the system by a call other than the one that began the link.
NTS List all NTS messages that have not been forwarded.
R # Read message #.
RN Read all new messages addressed to you.
SP call @ route Send a message to <call> at <route>.
SB name @ route Send a general interest bulletin of <name> via <route>.
SB name @ route $bid Send a general interest bulletin of <name> via <route> using <bid>. Note: A bid will be assigned by the system if not entered by the user.
ST id1 @ id2 Send NTS traffic using current conventions.
T Signal the Sysop for talk.
V Read system version number.
CANCEL # Cancels message # if you are the originator.

There is no Kill command. Messages that have been read or forwarded will be deleted automatically from the system after 24 hours. Unread messages will be deleted automatically after 21 days. Bulletins will remain until the originator or the Sysop cancels them.

Even though APLINK is considered primarily an AMTOR system by most hams, I think that you will find that most of the activity really occurs on the PACKET side. AMTOR is used primarily for forwarding the messages across the country and for the entering/retrieving messages by someone who can't get to the MBO by packet.

HINTS and KINKS for AMTOR OPERATION on APLINK

Jim Jennings , KE5HE @ KE5HE.TX.USA.NA

Here are some hints to make your use of any APLINK a more pleasurable and efficient experience:

Equipment setup:

1. Use AUTOLOGON (Set WRU=ON and AAB/WRU text to 'QRA CALL + ?'). This will enable the AUTOLOGIN procedure, after the first time.
2. Use 2125/2295 MARK/SPACE tones (170 Hertz shift) on transmit and receive.
3. Set your rig to within 20 Hertz of the MBO frequency. If you have a scope use it. Using LEDS accurately for tuning takes practice. Since most MBO's use sharp filters on receive, precise tuning is important. Once you have found the proper frequency, you might want to store it in memory for later use.
4. Be certain that your transceiver transmits and receives on the same frequency. This can be a problem.
5. Use 500 Hertz receive filters if your rig is equipped with them.

Procedures for connection:

1. Select a clear channel on the optimum band for time of day and distance (avoid 20 meters when possible).
2. Listen with SSB filter first, then move to a clear channel and use sharp filters.
3. Seize the LINK and send a letter 'A' followed by RETURN when necessary to abort the list of a long file.
4. LOGOFF and LINK on different frequency or band if throughput is less than 60 percent.
5. LINKDOWN if QRM persists. Move to a different frequency and RELINK.
6. Try to keep your LINKS to 20 minutes or less on busy MBO's when reading bulletins and help files. Someone may be trying to forward, leave, or pickup traffic.

Digipeating

Harry Ridenour - N0CCW

From the very earliest days of packet, it has been accepted that digipeating was (and is) one of the most inefficient and least desirable ways to communicate from one point to the next available on packet. This is because digipeating is the most prone to collisions and most likely to require retries from the lack of end to end acknowledgment. Digipeating was used extensively in the early days of packet, because we had little else. It was the only bridge to get us from where there were NO nodes or networking capabilities to some other location. The problem today is that digipeating was NEVER intended as a LONG TERM networking solution.

First, the term (Digipeater) and the action of (Digipeating) are NOT always synonymous. Many people use the term "digipeater" in a generic sense to cover ALL packet nodes. For this article, I will use digipeating to mean a specific act and the result of specific chosen operating parameters from the TNC. Connecting TO a user (VIA) a digipeater is a great deal different than connecting (TO) a user after having first Connected to a NODE.

Today we have the choice of several networking packages (NetRom, TheNet, TexNet, Rose Code, TCP/IP, KANODES, etc.) and yet we continue to see people use digipeating to establish connects. WHY?

I think it has to do with an educational problem. Most packet radio books STILL refer to digipeating with little mention of "networking". They tell the newcomer to use "via something", so he does—or a novice operator sees someone else using a digipeater, so he does. Most of the books don't explain the differences, the impact, or the disastrous effects of what can occur when some users are digipeating and other users are connecting to and thru the same network node.

What's wrong with digipeating? Well, primarily it has to do with two problems. The first problem is that a simple digipeat path between two users creates FOUR paths where a collision, a loss of data, or an error can occur which will require the retransmission of the all the data transmitted (see Figure A).

Figure A - Digipeater

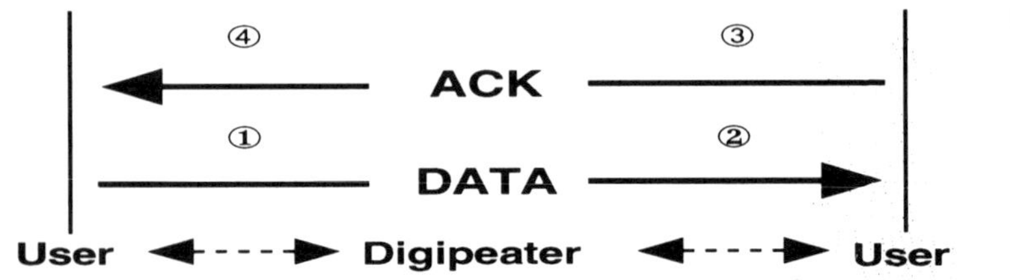

The second problem is that digipeaters are "dumb". Digipeaters have few programmable options like: DWAIT, FRACK, PPERSISTANCE, SLOTTIME, and a lot of other important settings. Digipeaters work on the simple principle of "as soon as the frequency is clear - I will send whatever I have, regardless of channel activity." This simple statement, coupled with the fact that most of the default parameters on TNCs are not set correctly, will virtually guarantee a collision with anyone else on frequency when digipeating.

On the other hand, with a network node there are only TWO possibilities of a collision or a retry for a given packet or packets (see Figure B).

Figure B - Network

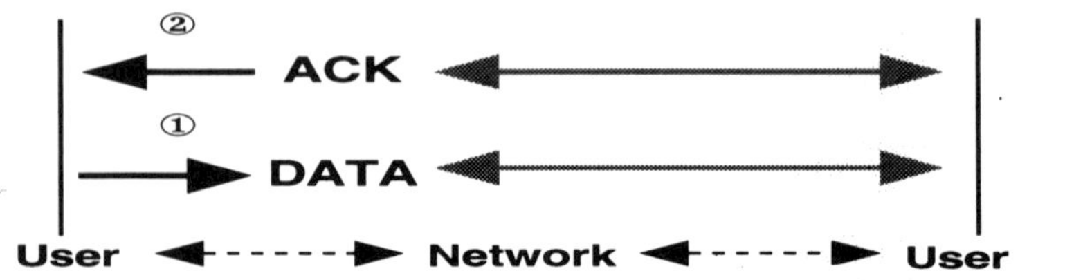

This reduces the chances of errors by 50% and increases the channel throughput, thus allowing more users more access to the same facility without retrying out. When using a digipeater or digipeaters to create a connection, any loss of data, including acknowledgments, requires the COMPLETE retransmission of all data packets from the user to the distant station. With network nodes, only the specific data lost between adjacent nodes need be retransmitted.

The advantage of using a network node over a digipeater is that the network node is programmable to insure better channel utilization and minimize collisions and retries with parameters like DWAIT, FRACK, PPERSISTANCE, SLOTTIME, and a lot more settings. One thing to remember is that most of the parameters designed to correct for collisions, only take effect AFTER the first loss of data has occurred! This doesn't help a digipeater; it has no "memory" of past events!!

Why is the digipeat function often left "turned on" with the newer network nodes? Why do we still have dedicated digipeaters? On the first question, it makes it possible for the "newcomer" to find his way onto the system and then you never know when you might really need to use digipeating in an emergency. On the second item, there are specific needs that certain digipeaters fill admirably. A very good example is SATN(X),WA5FSR-1, in San Antonio, which is a dual band, dual port digipeater providing instant access from 145.01 to 223.58, either as a digipeater or as a Rose node.

You may have heard digipeating is "faster" than using the networking capabilities. This is a half-truth. IF ALL packets get passed with NO retries, it's faster. If there is even ONE retry, it is slower. If there is more than one node, digipeater, or user on a frequency, then there are going to be retries. If there are ANY retries, the networking scheme will get through "faster". If there are four users, two using digipeating and two using network connects, the two using network connects will almost always succeed without retrying out over the digipeater users, and because there are digipeater users, the overall number of retries WILL be significantly higher FOR ALL INVOLVED. Hence those using digipeating are CAUSING unnecessary retries for EVERYONE else. It can be shown mathematically that connects over a three digipeater path have less than a 30% chance of completing a QSO. A three node network path has better than a 98% chance!!!

Reconsider the next time you initiate a connect. How about using the networking tools available and reduce the channel loading FOR EVERYONE and don't digipeat. If there are other choices available—NEVER digipeat over a node where network connects are operating. Avoid multiple digipeater paths like the plague. In this day and age there IS NO justification for multiple digipeater paths being used by users or BBSs to forward mail or to service users. It can do nothing but create havoc and hard feelings towards everyone concerned.

Prioritized Acknowledgment: An Overview

PriACK is available from TAPR in version 1.1.8 firmware.

The Idea

The idea behind the prioritized acknowledgment (PriACK) protocol is quite simple. Acknowledgments (ACKs) get priority access to the channel so that time is not wasted retrying packets that have already been correctly copied by the remote TNC. In normal AX.25 practice, for a variety of reasons, the ACK is often not received within the time limit defined by the FRACK (T1) protocol timer. In fact, because "channel access" methods are not clearly defined in the AX.25 Level Two Version 2.0 protocol document, the present protocol is more likely to synchronize collisions with acknowledgment packets than with any other type of packet!

The Problem

When a channel becomes clear, if your TNC has something to send, it will immediately send it. Unfortunately, if anyone else's TNC has something to send, it will also "jump on" the channel as soon as it becomes clear. The result is collisions and retries. Only after a collision does the pre-sent protocol suggest using random backoff. Thus, an acknowledgment to a just-received packet is almost assured of a collision-and-retry sequence, even though the TNC correctly received the packet. This will usually cause the TNC which is awaiting the ACK to time out and re-send the data.

To this collision synchronization mechanism, AX.25 Level Two Version 2.0 adds a propensity to cause even ACKs which are not from hidden terminals, and thus less susceptible to collision, to be delayed beyond even generous FRACK timer settings when the channel gets busy. Once FRACK times out, even if the ACK finally makes it through before the retry is sent, the original packet is retried anyway. This wastes a lot of time which could be better used clearing the channel of some of the legitimate offered load. This feature of AX.25 Level Two protocol accounts for much of the abysmal performance of the currently popular NETROM and THENET nodes when used with omnidirectional antennas systems.

How It Works

PriACK protocol avoids the above problems by giving ACKs priority access to the channel. It does this in such a way that even ACKs coming from hidden terminals are usually protected from collision. Standard AX.25 protocol gives a limited version of this priority access only to digipeated frames through the DWAIT timer. Ack prioritization works with slotted channel access in the following way:

1. Response frames (ACKs) are always sent immediately upon receipt of a valid packet. Ultimately, not even DCD will be checked for sending an ACK. However, in this release DCD will still hold an ACK off the channel.

2. Stations queued up to access the channel but waiting for a channel busy condition (DCD true) to clear, will start a slotted access procedure only after enough time for a response frame to clear the channel has transpired (ACKTIME). This is true even if the TNC waiting to use the channel cannot hear the ACK being sent.

3. Slot time windows (DEADTIME) are set large enough to ensure the local TNC will be able to accurately determine if another detectable station has started transmitting in any slot preceding the slot selected by the local TNC. This prevents two TNCs which have selected adjacent slots from colliding.

As you can see, under this protocol there will never be a condition where an ACK is delayed from being sent beyond the FRACK timer limitation. However, the FRACK timer is still active and must be set to a value that is long enough to allow time for (MAXFRAME * PACLEN) +ACKWAIT to expire before FRACK does. This time will depend on the radio and radio channel data rate in use. The TNC knows that if it doesn't see the ACK immediately when expected, it is probably never going to see it. (See discussion of new parameters below for definition of ACKWAIT.)

Forcing a channel access delay for all stations on the channel which are waiting for a chance to transmit allows ACKs from hidden terminals to get back to the expecting station. This clears that traffic from the offered channel load. If the packet was indeed copied and ACKed, further retries of the same information will not be necessary.

Multi-Connect Survival Guide

Why Multi-Connect?

Multiple connection capability is a powerful addition to your TNC 2. It is very useful for traffic net operation, multi-user bulletin boards, path checking and so forth.

What is Multi-Connect?

Multiple connection operation is not the same as multi-way operation. With multiple connect, you may establish several point-to-point "links" with various stations. Multi-way, which is not available, would enable multiple stations to be simultaneously interconnected to each other, with each station seeing all data passed from any station in the group, error free. Multiple connection operation is another step on the road to proper networking, and networking should eventually allow multi-way operation.

What commands set my TNC 2 to normal operation?

Your TNC 2 defaults the multi-connect-related commands to the following parameters:

CONPERM	OFF	STREAMCAll	OFF
STREAMDouBLe	OFF	STREAMSWitch	|
USERS	1		

This sets up your TNC 2 to act just like a "normal" TNC does that doesn't support multiple connections. The key to obtaining this traditional operation is to set USERS 1.

How Do I Invoke Multi-Connect?

If USERS is not 1, you are telling the TNC to allow multiple connections to your stations from other stations. In addition, TRANSPARENT mode will operate differently, in that incoming data will be prefixed with the current STREAMSWitch character and identifier (such as " | A"). The STREAMSWitch character, defaulted to " | ", should be set to a character you won't normally use. Note that this character may be set to a hex value between $80 and $FF. This may allow you to use 8-bit characters (AWLEN 8) if your terminal or computer is capable of generating such "characters." This could help prevent confusion in interpreting incoming data from other stations if they happen to send data that includes your selected STREAMSWitch character.

Although not foolproof, enabling STREAMDBL may also help in sorting out STREAMSWitch characters included in the received data from a valid stream switch generated by your TNC 2. STREAMCAll should be especially helpful when manually operating a station in which you allow multiple connections.

When in CONVERSE mode, you may switch streams by entering the STREAMSWitch character (default " | "), followed by a stream identifier ("A" through "J"), followed by the data you wish to send to the station on that stream. See the example in the description of STREAMCAll for an illustration of this.

If all this seems a bit confusing, don't worry, it is! The only way to really understand multipleconnect operation is to try it! Once you have your TNC 2 on-the-air, and have gained confidence in "normal" packet operation, try setting USERS 2 and get a couple of friends to connect to you. Play with the commands (you can't hurt the TNC by issuing commands to it!) and see their effects.

Happy packeting!

Conference Bridge Commands

/? or **/HELP** Display Help List (Command List).

/WHO **Display All Users and their Channel Numbers.**

/CHANNEL n **Switch to Channel n (0-255)**

/INVITE user **Invite user (callsign) to join Channel.**

/WRITE user or **/MSG user**

Send a message to user only.

/QUIT or /EXIT or **/BYE** Exit.

<u>Note</u>: Anything sent that is not a command will be sent to all users on the same channel. All stations ACKed individually. All commands may be abbreviated to one-letter (following the /).

APLINK PACKET Channel Commands

B Logoff the BBS.
Just disconnecting does the same thing.

H Sends Help file.

I Sends the Information file.

L Lists all messages.

L # Lists all files from message #.

LB List all bulletins.

L< call Lists all messages to <call>.
If <call> is omitted it lists all messages from you.

L> call Lists all messages to <call>.
If <call> is omitted it lists all messages to you.

LH List all help files.

LR List all users who have signed into the system in the last 24 hours.

LOGIN call Logs you into the system by a call other than the one that began the link.

NTS List all NTS messages that have not been forwarded.

R # Read message #.

RN Read all new messages addressed to you.

SP call @ route Send a message to <call> at <route>.

SB name @ route Send a general interest bulletin of <name> via <route>.

SB name @ route $bid Send a general interest bulletin of <name> via <route> using <bid>. Note: A bid will be assigned by the system if not entered by the user.

ST id1 @ id2 Send NTS traffic using current conventions.

T Signal the Sysop for talk.

V Read system version number.

CANCEL # Cancels message # if you are the originator.

Net/Rom Command Summary

A Net/Rom node provides the normal functions of an ordinary AX.25 digipeater, plus a set of sophisticated higher-level networking capabilities. A user may connect to a nearby Net/Rom node; display a directory of other known network nodes; establish a transport-layer circuit to another node; and call CQ or connect to another station in the vicinity of the remote node.

A Net/Rom node is connected to by using an ordinary connect request to the stations call or alias.

Commands :

? - Anything other than a Net/Rom command will return a list of Net/Rom commands.

CONNECT NODE - requests a circuit to another node or user via the network.

CQ [text message] - broadcasts a short message from the node the user is connected to.

NODES -displays the list of known nodes from the node you are connected to.

ROUTES - displays the neighbor list of the node's routing table.

USERS - displays a summary of who is using the node.

KANode Commands

ABORT	Stop a connection in progress
B(ye)	Node will disconnect
C(onnect) **call**	CONNECT TO callsign
C call S(tay)	Stay connected to node
J(heard)	Callsigns with daystamp
J S(hort)	Heard callsigns only
J L(ong)	Callsigns with daystamp & vias
N(odes)	Heard node callsigns daystamp
N S(hort)	Node callsigns only
N L(ong)	Node callsigns daystamp & vias
X(connect) **call**	Connect to callsign on other port
X call S(tay)	Stay connected when end DISC.

DX Cluster Commands

BYE	Disconnect from the node.
HELP or ?	help (displays this listing)
Show/DX	List the last 5 logged DX entries
Set/Name NAME	Would set your name to NAME
Set/QTH CITY	Would set your QTH as CITY
Show/Users	List currently connected users
Show/Users CALL	Show specific info on the user
Show/Config	Show the node configuration
Show/MUF wpx	Show MUF for wpx country
Show/Node	Show PacketCluster node list
Show/Notice	Show system notice information
Show/Sun wpx	Show sunrise/sunset for country
Show/Version	Show version of system
Show/WWV	Show WWV propagation info
Talk CALL	Enter talk mode with CALL
Talk CALL MSG	Send MSG to CALL
Conference	Enter Conference Mode
DELETE	Delete mail message
DIRECTORY	List active mail messages
READ	Read mail message
REPLY	Reply to the last-read message
SEND	Send mail message
Announce MSG	Make a general announcement
Announce/Local MSG	Announce to local node
DX FREQ CALL MSG	Post DX announcement that CALL is on FREQ with a short message.

It's probably best to get a feel for what is going on before you start to post.

Mailbox/BBS Command Summary

For help on a specific command, enter **H x** where **x** is the command you need help for. For example, **H R** will give complete help for the READ command. **? x** will give you a brief explanation of command **x**.

Summary excerpt of commands

@ BBS - Enter this symbol to indicate the BBS of the addressee, for proper forwarding of the message to its destination. The message, no matter to whom addressed, will be forwarded to the "@ BBS" location.

Bye - Log off the MailBox. Simply disconnecting has the same effect.

Info - Gives information on the equipment in operation.

K # Kills message number #.
KM Kills all messages that you have read.
KT # Kills an NTS message, generates return 'service message'

List Lists new messages since your last log-in.
LB Lists Bulletin messages.
LB Lists Traffic messages.
LM Lists all messages TO you.
L # Lists messages back to and including number #.
LL # Lists the last # messages.

Messages are listed in reverse order, newest to oldest. "Private" messages not to or from you will not be listed.

I CALL - Query the local white pages for all information on call.

N xxxx - Enter your first name into user data base.
NE - Toggle your "expert user" status.
NH xxxx - Enter your 'Home BBS'.

R # Read message number #.
RH # Read number #, showing all routing headers.
RM "Read Mine". Read all your unread messages.

S? xxxx @ yyy Send message type '**?**' to station '**xxxx**', at optional BBS '**yyy**'.
The MailBox will prompt for title and ask you to enter text. End text entry with a _ctrl-Z_.. "**?**" is an optional "type" of message.

They include:
- **B** - Bulletins. General information to all.
- **P** - Private. Only the addressee can read or list
- **T** - NTS Traffic

Talk - Chat with the Sysop.

ROSE net

Charles Burnett, AA5AV, College Station

The ROSE network is easy to use since one direct connect from a TNC will establish a link to another packeteer anywhere within the network. Primary and alternate routes are configured by the network managers. Information about the network is posted on each ROSE switch, so any packeteer can check his local switch for network help.

General Help

While DISCONNECTED, a connection using the ROSE X.25 network can be established by issuing a command like: CONNECT CALLSIGN-SSID VIA SWITCH,ADDRESS. Where CALLSIGN-SSID is the callsign that you wish to contact, SWITCH is the callsign of your local ROSE switch, and ADDRESS is the 6-digit network address of your contact. Note: Some tncs require you substitute the letter "I" for the numeral "1", and the letter "O" for the numeral "0" in the network address.

Applications

Switches may have extra applications available. These include:

INFO	displays network information
USERS	shows direct links available /who is using them
HEARD	displays what the switch has monitored recently

To use these, CONNECT to the application name at the network address of interest.

Calling CQ

To call CQ using the ROSE network, connect to CQ at the address of interest. You will receive a "remote station not responding" message. If someone sees your CQ, they will also see your network address and be able to connect back to you.

Excerpts from the ROSE X.25 Packet Switch User Manual

The Radio Amateur Telecommunications Society
206 North Vivyen Street Bergenfield, New Jersey 07621
United States of America 01-201-387-8896

Full documentation on ROSE is available on Compuserve HAMNET in Library 9.

Introduction to the ROSE X.25 Packet Switch

The ROSE X.25 Packet Switch is a connection-oriented, Open Systems Interconnection packet switch which conforms to the CCITT Recommendation X.25 and provides the user with a functionally rich network interface. The user interface to the ROSE X.25 Packet Switch has been designed with the average user in mind. Current users who are familiar with networking using digipeaters (C CALLSIGN VIA DIGI, DIGI) will find that we have continued this basic concept in the ROSE X.25 Switch user interface. The network will accept data from you and will notify you if there is a possibility that data has been lost. The network is 100% reliable unless you are otherwise notified. Users no longer need to know each step through the network to get to the desired destination. The network will handle all routing of connections as defined by the routing tables that the network manager has set up.

The only two things you need to know to make calls using the ROSE X.25 Packet Switch are the call sign of the switch local to you, and the network address (Area Code/Exchange in USA), of the point you want to exit the network. It's like knowing where the telephone is in your house and knowing the phone number (Telephone Network Address) of the person you are calling.

What ROSE Provides

The ROSE X.25 Packet Switch allows users to connect transparently through the Amateur Packet Network to a remote station without the worry of setting up connections on a step-by- step basis. All a user needs to do is connect through the local switch specifying the remote station's network address. To connect to a station with a ROSE X.25 Packet Switch you only need to know the call sign of your local switch and the network address (telephone area code and exchange) of the switch local to the station. For example, to connect to W2VY from anywhere in the USA: **C W2VY via *call_sign_of_your_local_switch*, 201478**

Disconnecting from a Station

You just disconnect like you normally would, if you are using a BBS send the "**BYE**" command, if you are talking to another person hit Control-C and type "D" at the "cmd:" prompt.

Information Bulletin

The switch contains a configurable information message which can be used for network information. To get this message you just need to connect to the switch and enter **RETURN**.

Local Digipeating

The ROSE X.25 Packet Switch will only digipeat frames with just ONE call sign (its own) in the "via" field of the AX.25 frame. The digipeat call sign is usually the call sign of the switch with a "-2" suffix, and the switch call sign is generally the station call sign with a "-3" suffix. In any case the suffix of the digipeater will be one less than the suffix of the switch. For the N2DSY-3 switch the following will work: "**C N2FWI V N2DSY-2**" But the following will be ignored: "**C N2FWI V N2DSY-2,KD6TH-5**"

Networking with ROSE

There is only one new concept for users to learn in order to use the advanced networking capabilities of the ROSE X.25 Packet Switch. Each switch has a unique, local, six-digit "address." This address is the telephone area code and the exchange of the location the switch is serving. This address is used to indicate to the network where the station you want to communicate with is located.

Local switching

Instead of digipeating you may want to use the advanced functionality of the switch to reduce channel overhead and increase the overall throughput. If you want to connect to another station that uses the same switch (N2DSY-3, address 201744 in this case) that you use, you can do this as shown; For example: "**C N2FWI Via N2DSY-3,201744**" Where you want to connect to N2FWI. In the preceding example you specified N2DSY-3 because that was where you entered the ROSE X.25 Network, and you specified 201744 because that was where you wanted to leave the ROSE X.25 Network.

This initially looks like a two hop digipeater connection, but in reality the ROSE X.25 Packet Switch gets into the picture and makes the connection more reliable. The switch will receive the connect request from you, establish a connection with you, and then attempt to establish a connection with N2FWI. This arrangement provides less congestion within the network because the acknowledgements are only between adjacent stations.

Call Progress Messages

When a connect request is accepted by the switch a message will be sent to notify you that the network is attempting to make the connection. This message is "Call being Setup" and once the Call being Setup connection is established end to end the switch will let you know by sending "**Call Completed to call @ address.**" "**call**" is the Call Completed to call @ address callsign of the station you asked for and "address" is the network address of that station.

Example:

cmd:*C N2FWI Via N2DSY-3,201744*
C N2FWI Via N2DSY-3,201744
Call being Setup
Call Completed to N2FWI-0 @ 3100201744

At this point normal communications can occur.

BBS Forwarding

To provide improved support of Bulletin Board mail forwarding, the ROSE Switch has an additional mode of operation for improved reliability. If instead of using the switch callsign (N2DSY-3 in these examples), the BBS uses the digipeater callsign (N2DSY-2 in these examples) the switch will enter the special mode. In this mode the network will not send any of the Call Progress messages and if there is a problem where a "*** Reset ***" would need to be sent, the connection will be cleared. This is to reduce the chances for message corruption or duplication.

Multi-switch networking

In order to connect to another station at a remote switch one must know the address for that switch. If KB7UV uses the switch "718956", a user would type: "**C KB7UV V N2DSY-3,718956**" to connect to him.

Entry and Exit Digipeaters

If the nearest ROSE X.25 Packet Switch is not local to you, you may need to include an extra digipeater address. For instance if K2MFF-2 is the digipeater you need to access the nearest switch (N2DSY-3), then you could use a connect command such as;
C WA6KJD Via K2MFF-2,N2DSY-3,619372.

A digipeater may also be needed at the exit point:
C WA6KJD Via N2DSY-3,619372,WA6KJD-2.

As well as Both:
C WA6KJD Via K2MFF-2,N2DSY-3,619372,WA6KJD-2

How to determine where a connection originated:
When monitoring a channel you will see the switch call sign preceded or followed by a six-digit number. This is the area code and exchange of a switch within a ROSE X.25 Network. Example:

If you were to see a monitored frame such as:
WA6KJD>W2VY,619372,N2DSY-3* <I>:
Carol says Hi!

Where the <I> indicated the transmission contains Information. This would indicate that the N2DSY-3 switch was transmitting information to W2VY on behalf of WA6KJD who is at Switch address 619372.

And if you saw:
W2VY>WA6KJD,N2DSY-3,619372 <I>:Hi Dad, back on HF yet?

Would be a frame sent by W2VY going to WA6KJD@619372.

TexNet Command Summary

Texas Packet Radio Society,P.O. Box 50238, Denton, Texas, 76206-0238

Introduction

TexNet is designed to be a dedicated, remote sited, multi-access, multi-resource network system. Users typically access network nodes using standard 1200 baud AFSK. This allows any user to have all the advantages of highspeed network operations without the requirement of purchasing additional station equipment beyond the normal TNC. The Texas TexNet network operates network links at 9600 baud.

TexNet Node Services

There are several services possible on each TexNet node. Use of any of the node services is selected by the Secondary Station Identification numbers (SSID). Example: C WR5C-4 would connect the user to the TexNet network service on WR5C. Although digipeating is discouraged, digipeating is supported by most nodes. Digipeater link limits vary by service and node programming, but in no case is more than 4 links in any frame's digipeater field accepted.

TexNet Local Services

-0 Digipeat operation

-0 SSID provides normal digipeater services. The node will recognize its call and alias. Any frames received by a node with more than 4 digipeaters or any frames longer than 256 characters are discarded. Use of the node as a wide area digipeater is not encouraged.

-2 & -3 Conference Bridge

A typical TexNet node supports two conference bridges. The user connects to a conference bridge by the usual connect sequence 'c WA5XYZ-2' or 'c WA5XYZ-3'. SSIDs -2 and -3 are two separate conference bridges. Each station wishing to take part in a roundtable needs to connect to the same bridge SSID. When at least 2 stations are connected and a packet is received by the bridge, the packet is duplicated. The duplicated packet is sent to all the stations connected to the bridge. Packets that fail to be acknowledged by each receiving station will be resent until the retry limit is exceeded. The one user command on the conference bridge is a Control-U. Control-U will return a listing of all the stations connected to that bridge at the time the request was received by the bridge.

-4 TexNet Network

Connecting to TexNet node's call with a SSID-4, TEXNET, or possibly the node's name, will allow the user to access the TexNet network.

TexNet Network Commands

A station connects to a TexNet node by using the -4 SSID. An example would be: c WR5C-4.
(The TexNet SSID for the WR5C NDALLAS TexNet node)

At the network prompt: NETWORK CMD? the user can issue network commands: HELP, BYE, LOCATIONS, CONNECT, CONNECT CQ, MESSAGE, WEATHER, or STATISTICS by typing the command in full or the first letter in upper or lower case.

An alternate way of accessing a node's TexNet network service is by using the node's name, instead of the node's call-4 SSID when connecting. Example: C DALLAS or C AUSTIN.

The NETWORK CMD? prompt is the base from which all TexNet network operations take place. After a connection is made over the network to a remote resource, the user will have to return back to the NETWORK CMD? prompt, before another TexNet network operation can take place. To return to the CMD? prompt, the user will typically issue a BYE command to the BBS or other service.

HELP or **?**
Help, h, H, or ? when typed at the network command prompt returns a menu listing the network commands.

BYE
BYE is used to disconnect the user from the TexNet network. Enter BYE at the NETWORK CMD ? prompt and the network will disconnect from the user. The only time a user should have to manually disconnect from the network is when the user has connected to another station that does not automatically disconnect (i.e. normal QSO). BBSs will automatically disconnect a user (when the user issues BYE on the BBS) and the network will place the user back at a NETWORK CMD ? prompt.

LOCATIONS
LOCATIONS returns a listing of all nodes on the network by name. The convention is to use meaningful city or place names for node names. Users should be able to understand where a node is located at first glance.

CONNECT
Connect is used to make connections to remote users or systems via the network. In this example, we want to connect to WB5DDP in the Austin area from W9DDD in the Dallas area.

At the NETWORK CMD ? prompt W9DDD sends:
c WB5DDP @ AUSTIN

TexNet at that point takes over; the connect request packet is sent to the node named AUSTIN and that node makes a connect attempt to WB5DDP.

If the network node, AUSTIN, makes the connection; W9DDD then sees from the Dallas network node: ***YOUR CONNECTION IS ESTABLISHED***

The receiving station, WB5DDP, will see :
******Connected to WA5LHS-4***
(WB5DDP's TNC is showing layer 2 connection and the callsign of the AUSTIN TexNet node)
Then WB5DDP will see : *** ***LINKED to W9DDD AT NDALLAS VIA TEXNET*** from the Austin TexNet node. At that point the 2

users are operating as if simplex and the network maintains the connection automatically.

The ***LINKED to message is currently compatible with a number of systems. This allows the remote system to recognize that they are connected to W9DDD (Layer 3) instead of the AUSTIN TexNet node (Layer 2), while maintaing local node identification for each packet.

If WB5DDP in Austin were out of range of the network node, but could be reached via a digipeater, then the connect request string would possibly read: ***c WB5DDP v DIGI @ AUSTIN***

Connections through at most 2 digipeaters are possible at either end of the network connection. Either aliases or callsigns can be used to make the connections.

During the connection sequence, the remote TexNet node will attempt 5 times to make a connection. If the connection is not made, the originating user will get: ***REMOTE USER NOT RESPONDING***

The user will be returned to the NETWORK CMD? prompt. A connection request to a TNC that is busy or a turned-off TNC will generate this response. If the remote user is currently busy using the network, the user will receive a Network Information Code (NIC) 13 and be returned to the NETWORK CMD? prompt. (see "Messages from aTexNet Node")

If the user wishes to have the network automatically disconnet on a network command failure, the user should use: C % CALL @ NODE. This connect command is useful for automated systems which are designed to get a DISCONNECT on failure and not a NETWORK CMD? prompt.

Connections through the secondary user port are possible. The user can force a connection request through a port other than the primary user port (port 1), if the user appends a comma and the physical channel number to the nodename.
Example: c WB5DDP @ ALAMO,2

The designations for the ports are: 0—Network backbone, 1—Primary User port, and 2—Alternate User port.

CONNECT CQ @ NODE
CONNECT CQ @ NODE is used to broadcast a CQ call over a remote node. In this example, we want to call CQ at the AUSTIN node from W9DDD in the Dallas area.

At the NETWORK CMD ? prompt, W9DDD sends : ***C CQ @ AUSTIN***

At this point the AUSTIN node transmits: ***CALLING CQ CQ CQ FROM W9DDD-0 @ NDALLAS***

The AUSTIN node will broadcast the above message 5 times at intervals of 8 seconds and then respond with : ***REMOTE USER NOT RESPONDING.***

This indicates that no users in AUSTIN connected back to your station in the 40 seconds you were broadcasting CQ.

If you wish to connect to a user that is calling CQ over a node, you must first connect to your local TexNet node and then connect to the user @ the user's TexNet node.

Example: WB5DDP in Austin saw W9DDD's CQ thru the AUSTIN node. WB5DDP would then connect to AUSTIN and issue : ***C W9DDD @ NDALLAS.***

At that point, the two stations would then be in contact. At the end of the QSO one of the stations must enter a disconnect at his TNC's command prompt, with the other station receiving the NETWORK CMD ? prompt.

IMPORTANT : If WB5DDP receives:
NETWORK INFO CODE 013 FROM NDALLAS

Then W9DDD's CQ timed out and W9DDD is currently busy doing something else on the network.

MESSAGE

MESSAGE connects the user to the network Packet Message System (PMS). The network automatically routes the user to the node assigned as the network PMS. The PMS is the system module that operates as a multi-connect mailbox as well as providing the emergency real time transfer and store functions for emergency traffic operations. The command structure once in the PMS system is a subset of the BBS command set. (see "PMS")

MESSAGE @ NODE

MESSAGE @ NODE connects the user to the Packet Message System (PMS) module at the designated network node indicated. At the prompt NETWORK CMD?, the user can type: M @ NWS. The network automatically connects the user to the PMS at NWS. See the network listing for alternate PMS locations. This command is used to get PMS's other than the network designated one.

WEATHER

WEATHER connects the user to the Packet Message System (PMS) that is designated as the weather server for the network. At the prompt NETWORK CMD?, the user can type: WEATHER. The network automatically routes the user to the weather server.

STATISTICS @ NODE

Statistics @ NODE returns information about accumulated node activity over a 24 hour period. Statistic counters are zeroed at midnight and when the node is reset. The user sends: S @ NDALLAS (to request stats at the North Dallas node).

Statistics from a node will return information regarding frame buffer usage, connect activity, and channel activity.

Frame Buffers have 350 byte buffers for received and transmitted frames. All channel counters must be multipled by 10 to get actual numbers. The RE-TXD counter is not multiplied.

S Y @ NODE accesses the stats from the previous day. Note that the time runs from MIDNIGHT to 00:00. Example: S Y @ AUSTIN.

This would return the previous day's statistics for AUSTIN in the above format.

ROUTE @ NODE

Route reads a node's routing table for the node listed in the @ field. The columns in the table (from left to right) are : node number of the node name listed on the far right, the primary via node, number of nodes on the primary path, secondary via node, number of nodes on the secondary path, and node name for the remote node.

The column indicating path length is weighted when the network link is established. Path length weighting will take place with 1200 baud network connections. Weighting is done to force the network to a higher speed path if one is available. Each 1200 baud link will add 3 to the number of "nodes" in the path length. In the alternate via path column, a node number of 000 indicates no secondary path. The path length indication then comes up as a default value of 75.

Packet Message Server

The connection through the network to the Packet Message Server (PMS) is automatic. The user will see: YOUR CONNECTION IS ESTABLISHED. The PMS will list any messages for your station. If no messages exist or the PMS lists all messages, the user will receive: PMS CMD ? (B,K,L,R,S) >

The commands are B(ye), K(ill), L(ist), R(ead) and S(end) all operate similarly to the now standard W0RLI command set.

Bye generates a disconnect from the PMS and returns the user to the NETWORK CMD ?.

Kill (number) will delete a message if you are the sender or receiver. Messages are automatically killed after 14 to 28 days on the PMS. KM will delete all messages addressed to your station.

Read (number) will display a message.

Send options are S callsign, SB callsign (for bulletins), ST callsign (for traffic), SP callsign (for private). The @BBS, used for forwarding, is not supported. The PMS is a local network message system only.

List will list messages stored on the PMS.
  L - list all messages entered since your last activity.
  LM - list all your messages.
  LB - list all bulletins.
  LL 5 - list last 5 messages. LL 24 is maximum.
  LW - list the latest NWS weather statements.
  LS - list all severe NWS weather statements.

A character the PMS doesn't recognize as a valid command will return: *BAD CHARACTER IN COMMAND LINE* or *BAD NUMBER*

A Read or Kill command of a message number not on the PMS will return: *MESSAGE NUMBER NOT ACTIVE.* A private message not addressed to you will return: *NOT YOUR MESSAGE.*

User Guide for TheNet+

Dave G8KBB @ GB7MXM.#36.GBR.EU [44.131.16.31] g8kbb.ampr.org, 7, Rowanhayes Close, Ipswich, IP2 9SX, England

This brief note is intended for users of TheNet X-1H, and explains the basic commands. Configuration and sysop features are not covered fully.

The switch provides the following user commands :

Connect	Info	Nodes
Routes	Users	Talk
CQ	BBS	Host
MHeard	Bye	DXcluster
IProute	ARP	QUIT

Not all commands may be available on every node as certain commands might have been disabled. If a command has been enabled, it will be displayed when you type an invalid command such as '?'. In addition, there are some commands that are available but are not displayed.

The main ones of interest are :

Links	Mode	Parms
Stats	IPAddress	DXCAlias
BBSAlias	HostAlias	

Connect

If the connect command is given on its own, then assuming that the sysop has set it up correctly, you will get connected to the local BBS. If you give another callsign, either of a local station or a node, the node will attempt to connect you to that station either by a level 4 connection or by downlinking. If you are downlinking, you may also specify digipeaters.

In either case, you get either a connected message or a message telling you of the failure to connect. If you enter any other command at this stage, the connection attempt will be aborted.

Info

This command gives information about the node as a combination of a message stored in the EPROM and a message entered by the Sysop.

Nodes

This command gives information about the distant nodes that this node thinks it can get to. With no parameter, it shows the alias and callsign of all the nodes except those staring with a '#' character. If a parameter of '*' is given, those 'hidden' nodes will also be shown. If a callsign or alias is given that the node does not know, it gives an error message. If the callsign or alias of a known node is given, the node gives details of the routes it knows about that lead to that destination. The display shows one option per line, each of which consists of the path quality, obsolescence count and port followed by the callsign of the neighbour. If any route is in use, a chevron is shown against the appropriate entry.

Routes

This command gives information about the neighbouring nodes that can be heard. For each neighbour, the display shows the port number, the callsign, the path quality and the number of nodes accessible through this neighbour. If a route has been 'locked' by the sysop, then a '!' character is shown after an entry. The sysop may have configured the node to display nodes as callsign or as alias:callsign.

Users

This shows who is using the node. It does not show other nodes that are using the node as a level 3 relay, nor does it show those users who have connected to the node but otherwise have done nothing.

The display shows the through connections, followed by those users who are connected to the switch and 'idle'. It also shows those users who are connected to the conferencing facility. The latter stations are shown connected to a destination called 'Talk', whilst in the case of connections, the two endpoints are shown. For connections, two symbols are used, '<-->' and '<..>'. The former is used for established connections whilst the latter is used for connections being established.

Talk

The Talk command allows a group of users to hold a conference call. It also allows a user to send a message to another user of the node provided that user is connected to the switch but is not patched through to another station and is not currently tyring to connect to another station.

A user enters the conference by giving the command 'talk'. He/she gets a message informing them of this and reminding them that the command to escape from the talk command is '/exit'. Any other users currently in the conference get a message from the node telling them of the callsign of the user who has joined them. At this point, every line sent by a user in the conference is copied to all other users in the conference, preceded by their callsign.

To exit from the conference, the command '/exit' is used. This causes a response message to be sent to the user, and at the same time all of those left in the conference get a message from the node telling them of the station who has left the conference. If you force a disconnect, the other stations are not told of your departure.

A string of text may be entered on the same line as the talk command when the command is given. If this is done, before the user is connected to the conference, that string of text is sent to all the other users of the node who appear in the 'user' list but are not connected to anything else. For example if GxABC were to type :

TALK GyXYZ, Hello fred can I have a chat - type TALK

then other users of the node (including presumably Fred, would get the message :

GxABC>> GyXYZ. Hello fred can I have a chat - type TALK

on their screens. The only exception to this is that sysops are not sent the message.

CQ

This command is used to broadcast a CQ message. In addition, the fact that you are calling CQ is indicated in the USER list. The callsign will be your own with a different SSID, and anyone else can connect to you by connecting to the callsign with the appropriate SSID.

The CQ remains 'primed' for a while, and if any other command is given to the node the CQ will be cancelled.

BBS

When you issue the BBS command, assuming that the sysop has configured it, you will be connected to the local BBS.

If you enter the command 'BBS ?', then the current setting of the BBS will be displayed.

Host

The HOST command operates just like the BBS command. It may have been disabled by the sysop, it may have been set to connect to the same station as the BBS, or it may have been set to connect to another host system.

If you enter the command 'HOST ?', then the current setting of the HOST will be displayed.

MHeard

If enabled, the heard list shows the last few stations heard. The number of entries is limited and set by the sysop so any stations not heard for a while may get pushed out of the list by others heard. Assuming that a station is not pushed out in this manner, the display shows the number of packets heard from that station since it appeared in the list and the time since it was last heard. The time is hours, minutes and seconds. The list also shows the port on which the station was heard (port 0 is the radio port), and if it hears IP frames or Net/Rom frames, it adds a note to show that the station is a node and/or a TCP/IP station.

If the list is long enough so that a station is not heard for 12 hours, it will get deleted anyway.

Links

The LINKS command shows the level 2 connections to the node. This is usually of academic interest, but I use it in testing. The display shows the links, one per line, with the two callsigns, the link state, the port number and the current number of retries.

Mode

The MODE command is a bit like the PARMS command. It shows a number of additional parameters. These are as follows as shown by example :

MODE THENET:G8KBB-5> 0 1800 6 3 2 20 0 600 2 900 1 31 0 1

with the following meanings :
0 Host mode protocol (0 = standard, 1 = DCD mode). 1800 = CWID period, Delay in seconds between CWID. 6 = CWID speed, 10's of msec per dot; 6 equals 20 wpm. 3 = Enable / disable nodes broadcasts mask. 2 = RS232 protocol, 0 = crosslink, 1,2 or 3 are KISS. 20 = TxDelay in 10's of milliseconds (Centiseconds ??). 0 = Full duplex control; 0 equals simplex. 600 = RS232 port nodes broadcast interval in seconds. 2 = Nodes broadcast algorithm port mask. 900 = Beacon period in seconds. 1 = 'connect' redirector, 0 is to HOST, 1 is to BBS. 31 = Each bit controls one of the 'user' help messages. 0 = This byte controls the broadcasting of 'hash' nodes. 1 = This byte enables / disables the extra alias operation. If you want additional details, ask the sysop for a copy of the overview guide.

Bye and Quit
These commands disconnects you from the node, closing the link. It says goodbye before disconnecting you if it has been so configured by the sysop. Quit does just the same as Bye does.

DXcluster
If there is a local DXcluster, this command may have been configured by the sysop to connect you to it. It therefore operates in a manner very similar to the BBS command.

Stats
The stats command gives lots of data about the node operation. A full description of the information is contained in the overview document.

IProute
This command is used by the sysop to configure the IP route table. It may also be used to display the router table.

Arp
This command is similar to the IProute command, but shows the Arp table. The Arp table provides a translation from Ip address to callsign.

IPaddress
This command is used to set or display the current node IP address.

BBSAlias, HostAlias, DXCAlias
These commands are used to set additional aliases for the node. It can be configured by the sysop to accept connect requests (uplinks) to the node callsign, the node alias, or the 3 aliases shown by these commands. When the node accepts a connection to one of these aliases, it will immediately invoke the BBS, DXC or HOST commands for you. The way this would normally be used is as follows. Suppose your local (for example) BBS was not accessible on the frequency that the node operates on. The BBS alias can be configured to provide easy access across other nodes to the BBS.

Full-Service Bulletin Boards - How to Better use Them

David Cheek - WA5MWD

Full-Service Bulletin Boards are provided for two purposes. First, some provide a place to find bulletins from the ARRL, AMSAT and local amateur groups. Second, they provide a person to person message delivery service. If you have a friend who operates packet radio, then you can have a full-service bulletin board deliver your message to your friend's home mailbox.

What is a home BBS? - This is the BBS you **EXPECT** to receive your packet messages at. In general it is **NOT** your home station. It is always a valid amateur call sign. Pick a W0RLI/ WA7MBL/AA4RE or other type of full-service bulletin board for your home. This will insure that you receive messages sent to you from distant BBSs. When you select a home full-service bulletin board by using the NH (Call) command, this Call is sent to the white pages server. Your friends can then find out which full-service bulletin board to send your messages to. The best choice for a home BBS is a 24 hour a day, full service BBS that you can reach quickly. Your local BBS operator will be happy to assist you in this subject. Just leave them a message on their BBS. It is important to note here that you should choose *ONE* BBS as your home full-service bulletin board, the one where you want all of your messages delivered. If you move or change your home BBS, you should then make sure that you update the information for your call in the White Pages database (see below). Just use the NH, NQ, and NZ commands on your new home BBS.

How is NTS traffic sent ? - NTS traffic are messages sent outside packet radio channels. They **MUST** conform to normal NTS guidelines, because it will be handled outside of packet radio systems, by voice and CW nets. Traffic **MUST** have a full name and address, including telephone number. It must have a normal NTS preamble. (this can be found in most ARRL operating texts, like the OPERATING MANUAL) Your message must also have **NO MORE** than 25 words. Contact your local NTS traffic handler for more details on how to format NTS traffic.

NTS traffic can be addressed in the following way:
TO (5 digit Zip) @ NTSXX where XX is the US postal service abbreviation for state. The title should include the city and state of Destination.

NOTICE: that you can use zip codes and NTSXX address forms to send private messages. The difference is that for private messages, you send the message by using the command: SP W5XYZ @ NTSXX. Example NTS traffic: ST 71000 @ NTSLA [for Louisiana traffic]

How to use the white pages? - Use the white pages to find the "Home BBS" of your amateur friends. White Pages was initially designed by Eric Williams, WD5CMU, of Richmond, California. The White Pages is a database of packet users showing their name, home BBS, QTH, and zip code. By accessing the database, you can determine if someone is active on packet radio and the callsign of the @ BBS to use when addressing them.

If your BBS is operating a local WP database, you can make an inquire using the "I" command. Enter I followed by the person's callsign. Example: I WD5IVD.

To get information from the national database you would send:

SP WP @ WD6CMU.CA
Title: WP Query

The message content is :

***callsign* QTH?** [one of these per line]
***callsign* QTH?** [one of these per line]
***callsign* QTH?** [one of these per line]
DE *your_callsign* @ *your_home_BBS*

You can make several inquiries in one message. You may get several replies for each inquiry. You can see that if someone uses their own call as the Home mailbox, we would not be gaining anything. The idea is to allow us to locate other amateurs, so we can send their messages to them. You have to know where they are to do that.

Foreign Country Third Party Traffic - NTS traffic is only permitted to a country that we have agreements with. The ARRL publishes a list of these countires. To send NTS traffic to listed foreign countries, the current approved NTS policy is to send all third party international traffic via the IATN International Traffic Net. Example: ST IATN @ NTSFL.

Making it quick - You can easily check for new items on the BBS with the following commands. Using them shows that you are a skilled operator.

LM for List Mine.
Shows all UNREAD personal messages.

LB for List Bulletins.
Shows all bulletins on the system.

LT for List nTs Traffic.
Shows NTS traffic waiting for delivery.

RM for Read Mine.
Reads all of your Unread messages.
Have your printer or disk capture on !

Sending Private Mail - All messages addressed to a callsign are private messages unless you specify a bulletin type. This is to keep the local mailbox from QRMing the local BBS frequency with long lists of messages that are not for you, and will just cause interference and delay if everyone reads them.

Hierarchical Addressing - The details are described in a paper presented at the 7 th ARRL computer networking conference. Hierarchical Addressing allows multiple address "domains" to be created, so that forwarding to distant locations can be simplified. The basic idea is that addressing is broken down to less detail as the address is read from left to right. An example is :

address @ BBScall.#local_area.state.country.continent
WD5IVD @ N5LDD.#NTX.TX.USA.NOAM

Reading from right to left in the @ BBS field, this means that N5LDD is located in Texas, which is in the USA, which is in North America. In this way, a message could be sent to a BBS from another BBS anywhere in the world, without any intermediate BBSs having to know anything about any BBS in Texas, or the U.S.A! Please notice that you just cannot make up these address parts! The state identifier is the old familiar two letter postal abbreviation. The country code is three letters from an international recommendation already being used for this purpose. The list of countries and continent designators is repeated in the proceedings of the 7th ARRL networking conference. The example just given can be sent by giving the following command.

Notice that you will not see the parts of the address which follow the N5LDD if you list the message on the BBS. They are included with the header and travel with it until the first non-compatible mailbox is reached. At that point, the address is shortened to the one address field on the left, in this case N5LDD.

Please remember that just like using a repeater, using a BBS is sharing a person's station, and a public frequency. Be considerate of others when using your local mailbox. Offer to help the BBS operator if you feel you like the service. Let them know when things just don't seem to be working right. I've often found users noticing things that I failed to catch while installing new software. Believe me, I appreciate getting constructive comments! I'm sure most other BBS operators do also.

Common BBS TO Fields

Jim Lill, WA2ZKD @ WB2PSI.#WNY.NY.USA.NOAM

Here's a listing of TO: keywords that I've gathered or have been sent by sysops. It is NOT intended to be an attempt at standardization nor does it mean you can't make up something new.

4SALE .... *DO NOT USE*
.... Use SALE instead (this is a SriLankan Callsign!)
ALL .... Use ONLY in last resort
AMSAT .... AMSAT (satellite etc)
AMTOR .... Amtor
ANT .... Antenna info
APPLE .... Apple // series (see MAC for Mac's!)
ARES .... Amateur Radio Emergency Service Info
ARRL .... ARRL bulletins
ATV .... Information on Amateur Television
CLASS .... Ham license class announcements
CLUB .... Club events, meetings, announcements
COMPTR .... Computer (non-specific)
CPM .... CP/M-80/86 Computer
DX .... DX reports and information
EXAM .... Ham license exam announcements
FCC .... FCC info
GRIPE .... Complaints (use with care!)
HAMFST .... Hamfest or flea market announcements

HELP	Requests for info, help, or questions
HUMOR	Funny quips, amusing stories, etc.
IBM	IBM (or compatible)
INFO	Other information
KANTRN	Kantronics Gear
KEPS	Keplarian Elements
MAC	Macintosh
MAP	Maps (Packet or otherwise)
MARS	Military Affiliate Radio Stations Info
MFJ	MFJ Gear
MOD	Request or info / equipment modifications
MODS	DO NOT use (see MOD)
NETWRK	Announcements about packet network
NTS	National Traffic System information
OSBORN	Osborne Computer
PACCOM	PacComm Gear
PART	Looking for Parts
PC	IBM/MS-DOS Computer, DO NOT USE (see IBM)
PIX	ASCII pictures
PROP	Propagation
PUBSVC	Announcements of public service events
PUZZLE	Puzzles
QSL	QSL card queries/reports
QST	Information
RACES	Races (also see ARES)
REQDB	Database Reqeusts (on some BBS's)
REQDIR	Directory Requests (on some BBS's)
REQFIL	File Requests (on some BBS's)
REQQTH	Callbook Requests (on some BBS's)
RTTYDX	VK2SG originated RTTY news.
SALE	Items for sale (MUST be Ham related)
SAT	All satellite information
SOFT	Software
SOLAR	Solar Flare etc
SPACE	SpaceNews originated by KB2BD
STOLEN	Stolen Ham Gear reports
SWL	SW Broadcast news
SYSOP	Provide/request information for Sysop
TAPR	Tucson Amateur Packet Radio
TANDY	Tandy and Radio Shack
TCPIP	TCP/IP (KA9Q)
TNX	Thank You notes
TRADE	Items for Trade (MUST be Ham related)
UNIX	UNIX Computer
USERS	Information for BBS users
WANT	Things wanted (MUST be Ham related)
WX	Weather Information (should be LOCAL in nature)

Formatting Messages for Transmission via the National Traffic System (NTS)

J. J. Leist KB5W @ WA5DVV

All messages passed through the National Traffic System should be prepared in a standard format to prevent garbling, misunderstanding, non-delivery, etc. Being in the standard form also permits handling of your message on all of the NTS facilities - not just packet. This may become necessary since packet does not yet reach all parts of the U.S. If you will follow these guidelines, you will ensure that your message has the best chance of arriving at its intended destination with little or no delay and in its original form.

Let's look at the whole picture of sending a message through a standard BBS, as an example. **BOLD** wording will represent what the user receives. The BBS> prompt is generic and will change from BBS to BBS. Everything else is supplied by the user. The example assumes you have already logged on to the BBS in question.

00 <u>MESSAGE TO - 'ST (5-digit Zip) @ NTS??'</u>
Where ?? is the US Postal Abbreviation for the State or Province. This tells the BBS how to route the message to it destination. 'qtc CITY ST' is used to help the BBS operator if the message has a problem and needs to be routed by hand. Example : ST 76201 @ NTSTX.

01 <u>MESSAGE NUMBER</u>
An identification number assigned at your station, unique to your station and independent of the number assigned by the BBS which accepts your message for relay.

02 <u>PRECEDENCE</u>
95% of all amateur traffic is ROUTINE (R). There are others which may be used:

Y = EMERGENCY This precedence is reserved for messages relating to a true emergency situation - one in which lives and/or significant property loss is involved. This ***IS NOT TO BE USED*** lightly or irresponsibly.

W = WELFARE This precedence is used on all messages which either request or report individual health and welfare status. Messages with this precedence should be processed ahead

00 **BBS>** ST ZIP @ NTS??

SUBJECT> qtc CITY ST

MESSAGE>

01	02	03	04	05	06	07	08
NR72	R	HXG	KB5W	16	Gautier MS	0210Z	OCT 17

09 JOHN LEE DOE
1234 SOMEWHERE ST
ANYWHERE TX 39999
214-555-1212

10 BT

11 THE TEXT GOES HERE BETWEEN THE BTS X ALL WORDS OR STAND ALONE LTRS ARE COUNTED

12 BT

13 SIGNATURE

AR

of all ROUTINE messages Remember, messages of this nature are sending information about loved ones which cannot be obtained by commercial means and should receive preferential treatment.

03 HANDLING INSTRUCTIONS

Special delivery instructions which you are requesting the relaying and/or delivering station to comply with. For example:

HXA followed_by_number - Collect landline delivery authorized by addressee within number of miles. (If no number, authorization is unlimited.)

HXB followed_by_number - Cancel message if not delivered within number of hours of filing time. Service back to originating station.

HXC - Report date and time of delivery (TOD) to originating station.

HXE - Delivering station get reply from addressee, originate message back.

HXF followed_by_number - hold delivery until specified date.

HXG - delivery by mail or landline toll call not required. If toll or other expense involved, cancel message and service originating station.

04 TEXT "CHECK"

The number of words used in the text of the message. This is the total number of words between the beginning and ending "BREAK" ("BT"). This includes the "X" . (NTS messages use *NO PUNCTUATION IN THE TEXT*.)

05 ORIGINATING STATION

The amateur call of the station actually originating the message.

06 CITY/STATE OF ORIGIN

Your QTH. Always use the location from which you sent the message.

07 TIME OF FILING

The time (in GMT/Z) which the message was given or written by you for transmission.

08 DATE OF FILING

The GMT/Z date on which the message was prepared.

09 DESTINATION ADDRESSEE

The person to whom the message is to be delivered. Include *COMPLETE* address with phone number. The more complete the address is, the better the chance the message has of being delivered.

10 FIRST BREAK

Indicates the start of the textual content.

11 TEXT

The information you wish to be conveyed. The letter "X" should be used for all punctuation.

12 LAST BREAK

Indicates the end of the textual content.

13 SIGNATURE

The identification of the sender. Special care should be taken with requests for welfare information to include the full name, address and phone number of the party requesting the information.

Conclusion

I hope this short explanation will help in creating your messages. By following these guidelines, you can be sure that your message is in the right form and that it will be handled as expeditiously as possible. If you have any further questions you should contact a local traffic handler for more information.

Handling third-party (non-amateur) traffic is a powerful public relations tool for amateur radio and should be exercised whenever the opportunity arises. The amateur community receives beneficial recognition whenever we perform this service for the non-amateur public.

How to Repair Packet Equipment

Will also work for other types of electronics.

STEP 1 Approach the ailing packet unit in a confident manner. This will give the unit the mistaken idea that you know something. It will also impress anyone who happens to be looking, and if the unit should suddenly start working, you will be credited with its repair. If this step fails to work, proceed to step 2.

STEP 2 Wave the manual at the unit. This will make it assume that you are at least familiar with the source of knowledge. Should this step fail to work, proceed to 3.

STEP 3 In a forcible manner, recite Ohm's Law to the instrument (CAUTION: before taking this step, refer to some reliable handbook to be sure of your knowledge of Ohm's Law). This will prove to the unit, beyond the shadow of a doubt, that you happen to know something. This is a drastic step and should be attempted only if the first two fail. If this step fails to work, proceed to step 4.

STEP 4 Jar the unit slightly. This may require anything from a three to six foot drop, preferably on a concrete floor. However, you must be careful with this step because, while jarring is an approved method of repair, you must not mar the floor or get the unit out of tune. Again, this is a very drastic step. If it fails, proceed to step 5.

STEP 5 Brandish a large screwdriver in a menacing manner. This will frighten the unit and demonstrate your knowledge of the deadly "short circuit" technique. Proceed to step 6.

STEP 6 Add a tube...even though the unit is solid state... This will prove that you are familiar with the node's design. This will confuse the unit and thereby increase your advantage. If this doesn't work, proceed to the most drastic and dangerous step. It is seldom needed and is a final resort if all else fails!!

STEP 7 T H I N K . . . !

Technical Terms

Addressing (TCP/IP)

Each Callsign gets assigned an IP (Internet Protocol) address. These are used as part of the networking scheme.

AFSK (Audio Frequency Shift Keying)

This is a mechanism for sending digital information over a radio. A signal 0 is sent using one tone while the signal 1 is sent using a different tone.

ASCII (American Standard Code for Information Interchange)

The American Standard Code for Information Interchange is a 7-bit code in which the numbers from 0 to 127 stand for text characters (including the alphabet, digits 0-9, punctuation, special characters and control characters). Each character consists of a defined combination of bits (0's and 1's). An ASCII file is a text file that does not contain formatting commands. It is used for transmitting text between computing devices.

Asynchronous Access

Accessing information in a delayed manner. The exchange of information does not require that both or all parties be present on the system at the same time to exchange information. Examples of asynchronous access are e-mail and new systems.

Asynchronous Protocol

A communications protocol that controls an asynchronous transmission. Example: ASCII, Kermit, and Xmodem.

Autoforward

Many of the bulletin board and mail server programs (BBS) are capable of passing messages to each other. System can be configured to autoforward message onto the next BBS when sending mail between BBSs.

Autorouting

This is a process by which a network node can pass traffic to another node via one or more intermediate nodes.

AX.25

This is the designation for the protocol used by TNCs to talk to one another. Based on AX.25 Version 2.2 as published by the ARRL. AX.25 is based on the X.25 commercial switching standard.

Backbone

The part of a network that handles the major traffic. It may interconnect multiple locations, and smaller networks may be attached to it. Internet is an example of a backbone system on a large scale. The backbone should be configured to optimize the data being transferred, rather than optimized for user stations. It is important for this optimization to have nothing but network nodes on the backbone. Backbones typically operate faster (9600bps+) than the local user frequency (1200bps) to help pass data in a more efficient manner.

Baud Rate

Baud is a measure of data flow. The technical definition is the switching speed of a line, which is the number of changes in the electrical state of the line per second. One baud indicates one transition of data. Baud rate is equivalent to bits per second at low speeds, such as, 300 baud is the same as 300 bps. However, at higher speeds, the bits per second is greater than the baud rate, because one baud can be made to represent more than one bit due to using techniques to enhance modulation. Example: A V.22bis modem provides 2,400 bps at 600 baud. It is an error to call a 2400 bps modem a 2400 baud modem. In packet radio, one character of text equals 8 bits of information. 8 bits of information requires 8 transitions of the modem tones. That means that there will be approximately 150 characters of data in one second of 1200 baud transmission. However, 150 characters is not seen in packet radio communication, due to the delay between sending packets, which can decrease significantly on very busy channels.

BBS (Bulletin Board System) or PBBS (Packet Bulletin Board System)

A computer system that functions as a centralized information source and message switching system for a particular interest group. Users connect to the bulletin board, review and leave messages as well as share information and programs with other users on the system. Bulletin boards may provide access, or doors, to other application programs (callbooks, HF ports, etc)

bps (Bits Per Second), bit rate

Used to measure the speed of data transfer in a communications system. The effective data rate is about 10% of the bit rate; therefore, 300 bps is equivalent to 30 characters per second. It would take a full minute to fill up a screen at 300 bps; 15 seconds at 1200 bps and about seven seconds at 2400 bps.

Beam Antenna

A directional antenna. The key features of a beam are 1) Directionality and 2) gain. In this case, gain means additional signal is sent in the direction the beam is directed.

Callbook Server

This is a network server whose function is to allow stations to access a callbook database. These servers are operated both stand-alone and as part of DX Packet Cluster.

Card or PC Card

A printed circuit card that is plugged into the inside of a computer to provide a feature, such as a serial port, that was not included with the original computer.

Carriage Return

On a mechanical typewriter the motion of pushing the carriage, which held the paper, to the left all the way until it stops is called a carriage return. When electric typewriters came into existence the key to the right hand end of the typewriter keys marked J, K, L, ;, ' was called the return key or carriage return key. On modern computer, the key is unmarked or labeled enter, return, or carriage return. The function of the carriage return key is to instruct the computer the end of a line has been reached.

Choke/Unchoke

See XON/XOFF

Circuit or Virtual Circuit

In many networks, a circuit is an assigned connection between two nodes. The circuit can between two nodes adjoining each other or may transit through other nodes to reach each other. This information concerning the route is stored in routing tables. The circuit exists until the destination user or the originating user/service disconnects, or until one of the two nodes decides that data cannot be sent any more.

Collision

This is an event, where a receiving station doesn't receive it's desired packet because another packet was generated by a different packet station in the same time frame as the desired packet and interference occurs. In this case, the sending station will wait a backofftime and the packet is retried or a special poll packet is generated. (See backoff, poll-packet, retry)

COM1 / COM2

The name assigned to serial port #1 or port #2 on a Personal Computer. COM ports are usually connected to a modem, mouse, and occasionally printers. LPT1 is used for printer support.

Command file

A text file saved on a disk that contains a list of commands to be executed automatically. Also called a script.

Commodore 64

Before everybody had their own computer, Commodore was one of the companies that cranked out a real low-cost computer. The Commodore 64 is the only one of those 1970s low budget computers that is still popular. Others include the TI99 and ZX81. The Commodore 64 is a computer in a keyboard. The entire computer guts are located inside the plastic case which looks like an ordinary keyboard. Enclosed in the box are 64K of memory (equivalent to 64000 characters of text) which is relatively small compared to modern PC compatibles, Macintosh, and Amiga models. The software for the Commodore 64 is usually built into accessories that plug into an expansion slot, or available on cassette tape. The Commodore 64, unlike the Macintosh, Apple 2, and PC compatible, often uses a normal television set for a video display. The Commodore only gets about 40 characters of text across the screen and about 16 lines of text. For packet radio, 80 characters is the standard line length, so 40 characters is a handicap.

Communications Act, 1934
Established the Federal Communications Commission (FCC) in 1934, the regulatory body for interstate and foreign telecommunications. Its mission is to provide high-quality services at reasonable cost to everyone in the U.S. on a nondiscriminatory basis.

Control character
On all keyboards which meet the ASCII standard there is a key marked as Ctl, Cntrl, Ctrl or Control. This key operates like the shift key, except that when that key is pressed the values associated with other keys are changed based on the control key not the shift key. When a normal alphabet key is pressed the control value of that key is sent, instead of the normal alphabet value. This control value is called the control character.

Conference Node, Chat Bridge, Crowd
These are names given to sites that support the ability to have more than one user chat together. Packet radio is typically a one to one communications, but with the help of a bridge, users can connect to the node and then communicate to any other user connected to the same node.

Connect time
The amount of time spent on-line with an information service.

Coverage
The area in which a signal can be heard is that signal's coverage. Usually coverage refers to both transmit and receive. If a station can both talk to and receive from a station in a particular area that station has coverage in that particular area. Wide area coverage means that the station can be heard and can hear a wide area. If a station can be heard in an area that it can't hear from then there are problems. Certain derogatory words have been used to describe those problems, like alligator (mouth, no ears).

CSMA (Carrier Sense, Multiple Access)
This is a system of packet operation that requires that all stations wait for the channel to be quite before transmitting. Most amateur radio packet uses CSMA.

Data Bits
When characters are being transmitted they are coded in "chunks" of seven or eight bits.

Data Transfer
The movement of data within a system or network. A transfer is actually a copy function since the data is not automatically erased at the source.

DCE (Data Communications Equipment)
A DCE is a communications device that establishes, maintains, and terminates a session on a communications network. A telephone modem or TNC is an example of a DCE. A computer is a DTE (Data Terminal Equipment).

Dedicated port
This is a port designated for a specific purpose with only one other station on the frequency, usually a tie-in to a server or other network hardware.

Digipeater (Digital Repeater)
All TNCs can be configured for relaying messages on a single frequency. Originally this option was meant to be used until such time as networks were built. Due to the way digipeaters operate, most packet organizations recommend against digipeating in any form except in emergencies. Users should use network whenever possible.

Diode Matrix
TNCs running ROSE, NET/ROM, TheNET network firmware can communicate to each other using EIA-232 (serial ports). If more than two TNCs are used a diode matrix is required.

Dip Switches
These are the tiny configuration switches on many modems and serial port cards.

Disk Drive
A disk drive is a computer accessory that stores data in the form of magnetic impulses on a flat media that is much like an extremely high quality magnetic recording tape. The media is on a spinning platter, like a phonograph, and is read and written by a tape head that moves across the media. Floppy media is commonly used for transfer of information between computers and for storage of seldom used programs.

DNS (Domain Name Server)
This is a host - or multiple hosts - that can resolve a Callsign into an IP address. KF5MG resolves to 44.28.0.14. This system is used in TCP/IP based networks.

DOVE
An OSCAR satellite (OSCAR 17) whose full name is Digital Orbiting Voice Encoder.

Download
To transfer information (files) from a host computer 'down to' your computer.

DTE (Data Terminating Equipment)
A communications device that is the source or destination of signals on a network. A terminal or computer is a good example of a DTE. DCE to DTE communications between a modem and a computer is typically used in data communications.

DTR (Data Terminal Ready)
DTR is a control signal that the computer can send to a modem to cause the modem to hang up the phone under program control.

DX Packet Cluster or Packet Cluster

A server used by HF operators to pass information about DX contacts. This software, originally written by AKIA, also provides additional HF-related information. DX Packet Cluster networks (multiple DX Packet Clusters, interconnected) share contact information (called DX Spots) in real time. When a HF station locates a rare DX contact they post the location of the DX station for all those connected to the DX Packet Cluster to operate if they wish.

Dynamic Routing

In a network where redundancy exists in the backbones from one city to another, some types of network software allow for the network to recover automatically from a backbone hardware failure by rerouting traffic through the redundant link. This is called 'dynamic routing' as it can adjust dynamically to a changing network. (See ROSE, TCP/IP, TexNet).

EOC (Emergency Operations Center)

This is a term used by some state governments for a state or county government owned facility at which emergency services radio equipment and other gear is clustered. Some E.O.C. facilities are located in bomb shelters, others in buildings as part of a state office complex. A good packet network would include an on line TNC or BBS at every E.O.C. in the network's region.

EIA-232 (aka RS-232)

Name of the cable standard used in attaching serial ports to other serial ports, modems, or printers. EIA-232 is a standard 25-wire electrical interface between a computer and a peripheral device (i.e., modem, mouse, drawing tablet, or printer). It is an EIA (Electronic Industry Association) standard for serial transmission that uses a 25-pin DB-25 or 9-pin DB-9 connector. A serial modem connection uses the EIA-232 standard. This is also known as V.24.

EIA-232 serial cables:

25 pin	9-pin equivalent	25 pin	9-pin equivalent
1 - Protective ground		14 - Secondary transmitted data	
2 - Transmitted data	3	15 - DCE transmitter signal element timing	
3 - Received data	2	16 - Secondary received data	
4 - Request to send (RTS)	7	17 - Receiver signal element timing	
5 - Clear to send (CTS)	8	18 -	
6 - Data set ready (DSR)	6	19 - Secondary request to send	
7 - Signal ground	5	20 - Data terminal ready (DTR)	4
8 - Carrier detect (DCD)	1	21 - Signal quality detector	
9 - + voltage		22 - Ring indicator	9
10 - - voltage		23 - Data signal rate selector	
11 -		24 - DTE transmitter signal element timing	
12 - Secondary carrier detect		25 -	
13 - Secondary clear to send			

Electronic Mail (E-mail)

Transmitting memos and messages over a network. Electronic Mail Systems or E-Mail is implemented in mainframe, mini, and personal computer LANs. Users can send mail to a single recipient or broadcast it to multiple users on the system. To send or receive a message usually requires an ID number or address. On packet radio BBSs user callsign and the callsign and hierarchical address of the BBS is used.

EPROM (Erasable Programmable Read Only Memory)

This is an integrated circuit (IC) which is used in computers, including TNCs, to permanently hold a computer program. In TNCs all of the program is located in one EPROM.

Exposed Receiver Syndrome

This is a condition where a packet station is unable to transmit due to the fact that it perceives the channel as being active continuously. This can be caused by Hidden Transmitter Syndrome and is often the case when a node is located on a high hill with surrounding metro areas. It can also be caused when noise levels on a channel increase to such that TNCs without true Data Carrier Detect (DCD) will show activity when it hears noise.

False Route

In a network using NET/ROM or TheNET software the node routing is generated automatically by the nodes themselves. If improperly managed, it is quite possible for routing to be discovered and used by the nodes such that DX paths are treated as real paths. In this case, a route may be created in the routing table that depends on a 'lift' (propagation enhancement) condition. When the lift goes away, the nodes will be helplessly trying the 'false route'. This condition is preventable by manually controlling the route tables to specify valid routes to neighbor nodes. This situation cannot occur with ROSE or TCP/IP software, since all neighbor nodes and routing information are defined by high-order routing logic or done manually.

FAX or Facsimile

A device to communicate a printed page between remote locations. Fax machines scan a paper form and convert its image into a code for transmission over the telephone system. The receiving machine reconverts the codes and prints a facsimile of the original. A fax machine is made up of a scanner, printer and fax modem. Fax Groups 1 and 2 transmit a single page in analog code at six and three minutes respectively. Group 3 Fax transmits one page of information up to 9,600 bps in less than one minute, at a resolution of 203x98 dpi in standard mode and 203x196 dpi in fine mode.

F.E.M.A. (Federal Emergency Management Agency)

FEMA is a government agency responsible for disaster recovery. They work closely with many regional packet groups in order to establish regional networking for emergency use.

FINGER (TCP/IP Command)
Command will get you information on a user or topic from an IP host. Finger@host will get you a list of users and topics. Finger user@host will have the info sent to you. Information does NOT get sent as a mail item.

Flame
Slang for communicating emotionally and/or excessively via electronic mail.

Floppy Disk
See Disk Drive

Foreground/Background Tasks
The priority given to programs in a multi-tasking environment, such as UNIX. Users can select operations to run in the background or foreground. In addition, the user can interrupt a process (i.e., mail, news) in order to do another task and then return to the task. This would move the foreground task to the background until the user called the halted background task to the foreground again.

Forward
See Autoforward.

Forward File
This is the disk file on a packet bulletin board system (PBBS) that is responsible for directing the autoforwarding operation. By making entries in this file, the PBBS system may select what packet paths are used to each PBBS that is forwarded to, when each operation is performed and what traffic is sent during each piece of the forwarding operation. (See autoforward)

FRACK (FRame ACKnowledge delay)
This is the time after a packet is transmitted by a TNC before the TNC decides that a frame acknowledge is not going to occur. At that point the TNC performs backoff(some TNCs + TCP/IP) and a retry. (See backoff, retry).

Frame
A frame, or packet frame, is a single packet. Several frames may be sent in one transmission but each has it's own address, start, and stop. Frames are also called 'packets'.

FTP (File Transfer Protocol)
A TCP/IP protocol that is used to log onto the network, list directories and copy files. Files can be sent as either ASCII or Binary Files.

Full Duplex
A setting in modem software that is used when characters are echoed from a remote system back to your screen. Most packet satellite stations use full duplex, while most terrestrial packet stations use half duplex. (See also: Half-Duplex).

Gateway

A computer that connects two different types of communications networks together or allows interconnection to another information system.

G8BPQ Code, BPQ

John Wiseman, G8BPQ, developed a TSR (terminate stay resident) program for the IBM PC and compatibles that would imitate NET/ROM and allow node access for a program that runs on the PC. This program simulates the NET/ROM node functionality and allows routing from a NET/ROM or TheNET system directly to the PBBS or other program running on the PC. Unlike a standard node, which can only handle one radio per TNC, the G8BPQ program may direct traffic in and out of several radios by using KISS TNCs or other TNC/modem cards.

Half Duplex

A setting in modem software which is used when the remote system is not echoing characters back to your screen. Most terrestrial packet stations use half duplex, while packet satellite stations use full duplex. (See also: Full-Duplex).

Handshaking, Modem

The introductory tones and data exchanged between telephone modems before normal data communication begins. These tones and data will occur between the answering of the phone and the appearance of the '***CONNECTED' message on your screen. Different speed modems have different handshaking sequences. Current Packet AFSK and FSK modems do not use handshaking tones. Future Packet Modems will probably support some form of modem handshaking.

Hard Disk

See Disk Drive

Host Computer

A computer that manages information for many users. A host computer may be a mainframe, minicomputer, or another microcomputer.

Host Mode

WB8DED created a software package for the TAPR TNC-1 that is called Host Mode. This package was later created for the TNC-2. Some BBS programs took advantage of the command language in Host Mode to control the TNC and to allow multiple users to connect to the BBS at the same time.

Hidden Transmitter Syndrome, Hidden Transmitter

This describes a condition where throughput is drastically reduced to well below the specified baud rate because a single station is able to hear two or more stations that can't hear each other.

Hidden Transmitter Syndrome-Free

By making sure that every radio/TNC on a frequency can hear every other radio/TNC, most of the collision problems and inherent loss of throughput can be eliminated. At this point, backoff becomes effective and the performance of the system of radio/TNCs may be predicted more accurately. The only remaining problems occur when radio dead time due to slow transmit receive switching is excessive. Also backoff must be used if there are more than two radio/TNC sets on frequency. Bit-Regenerating Repeater is one example of a HTS-Free System.

Internet

Internet is a large network made up of a number of smaller networks. Regional networks have access to this international network. The only thing required in common between all computers and networks using the Internet is that they all use TCP/IP as a common networking standard.

KERMIT

An asynchronous communications protocol for personal computers developed at Columbia University and used in several public-domain communications programs. It is a slow protocol but is noted for its transmission accuracy over noisy telephone lines because it checks for errors in the commands it sends as well as the data.

KISS (Keep It Simple Stupid)

In the packet world, this usually refers to a program or mode that relates to TNCs in which the functionality of a TNC is entirely remote controlled by an external PC or other host. KISS mode is used by G8BPQ code and TCP/IP.

LAN (Local Area Network)

Regional packet groups define LANs differently. Typically, LANs are a smaller part of a larger network or a network of small size. LAN size is pretty much defined by who creates the network. Other types of Networks include MAN (Metro Area Network) and WAN (Wide Area Network).

Local Echo

A method of communication in which your modem or software displays data "locally" on your screen, without relying on the host computer to echo the characters back. (See Full-Duplex, Half-Duplex, Echo).

Locked Node

Net/Rom and TheNET nodes have the capacity to generate routing lists automatically based on parameters set in the node's RAM. The parameters specify default quality values to be assigned to routes to each neighbor, separately defined for neighbors. A locked node is where an individual node is given a specific quality and nearest neighbor route, but for which the duration is set to infinite. This locked node must be manually entered by a sysop (system operator), but is visible to anybody who wants to look for it. A locked node may be used to make sure that a node listing doesn't go away, to set a node in with a route that is different than which may be automatically generated, to set a node in with a quality which is less than that which is automatically generated, or to make a node not show up at all.

Locked Route

Net/Rom and TheNET nodes have the capacity to generate routing lists automatically based on parameters set in the node's RAM. The parameters specify default quality values to be assigned to routes to each neighbor. It is possible using the sysop's ROUTE command to manually set a route at a specified quality. This may be done so that the routes never change due to weather induced radio performance effects, due to accident or due to malicious intent by another party.

Log On

The sequence used at the beginning of a session to enter your name or ID number and password and begin using the system.

Log Off (^D / Control D/ Exit/ Quit / Bye)

Logoff is the command used to quit using the system and terminate the communications session. On Unix systems it is (Control-D) ^D, on most packet services it is quit or bye.

Macintosh

A personal computer which is a product of Apple Corporation. A Macintosh is a windows-oriented computer. Most programs available for amateur radio are written for IBM PCs so the Macintosh is not as popular for hams as the IBM PC compatible computer.

Mail Box / Mail Drop / PMS / Personal Message System

The computer storage assigned to a user for electronically transmitted mail. This storage can be on your local BBS that you access. Some TNCs provide Mail Box capability within the TNC. Mail Drop is what AEA calls this function. It is also called PMS or Personal Message System by PacComm.

Matrix

See Diode Matrix

Modem

A modem is a MOdulator-DEModulator device that adapts a terminal or computer to a telephone line or radio. This is a hardware unit which allows a computer to transmit or receive data. The modem may be external or internal to the computer. It converts the computer's digital pulses into frequencies within the audio range of the telephone or radio and converts them back into digital pulses at the receiving side. The effective data rate is about 10% of the bit rate; therefore, 300 bps is equivalent to 30 characters per second. It would take a full minute to fill up a screen at 300 bps; 15 seconds at 1200 bps and about seven seconds at 2400 bps.

Multi-Streaming / Multi-Connect

This is the process by which a user can connect to several stations at once.

Neighbor

In a network of nodes, the neighbor of a node is any node that is talked to directly.

NET (KA9Q TCP/IP)

NET is a version of TCP/IP written by Phil Karn, KA9Q. NET was Phil's original attempt at a TCP/IP program for amateur radio use.

Node

A node is an active element in a network. This can mean anything from a user station to a bulletin board. Traditionally a node in packet radio is an intelligent router of real time data, somewhat more intelligent than a digipeater but faster than a store-and-forward BBS.

Node Broadcasts

Each node sends a one way message out every half hour (setable) that tells it's neighbors what nodes are listed in the nodes table. The neighbor TheNET nodes interpret this information based on a factor called quality. (See Nodes Broadcasts which is part of Theory of Operation in the Resource Manual part of this Notebook)

NOS (TCP/IP)

Version of TCP/IP written by Phil Karn, KA9Q. NOS is written in 'Turbo C' and has been ported for DOS, UNIX, Macintosh, Amiga, and other systems with an ANSI C compiler. Source code is freely available. NOS is being updated all of the time by many hams. There are various flavors of NOS: JNOS, etc.

NOS (Network Operating System)

NOS is a program which is generally used to communicate using the TCP/IP protocols, but which actually is much more than just a program that does TCP/IP. NOS is the name used to describe dozens of different programs which are all similar and which all are used for TCP/IP, but which have huge differences in look and feel and features supported. Examples: Finger, Ping, SMTP, etc.

On-Line

Describes activities and equipment currently connected to and accessible by the remote computer. The term can also mean 'being electronically connected' to another computer.

OEM (Official Equipment Manufacturer)

This is a term used to describe a manufacturer that has purchased or been given a license to reproduce a product. All the commercial TAPR TNC-2 Clones were produced by TAPR OEM's.

Packet

A packet is a block of many characters (or bytes) which are sent together along with a few extra characters. The packet includes addressing information so that the receiving station knows that the packet is for it as well as who sent the packet. (See Frame)

Parameters

In node and BBS software there is a list of values used to configure options. These values are called parameters or PARMS.

PARMS

See Parameters

Path

The route between any two nodes. Same as line, channel, link, or circuit.

Pathname

A pathname identifies a particular file or directory uniquely. The cd (change directory) command is how directories are changed in UNIX and PCs. Either an absolute pathname or a relative pathname can be used. An absolute pathname starts at the / (root) and lists each directory to the destination. A / (slash) is used to separate each subdirectory.

Parity

A method of detecting errors in data transmission. Not used by most systems, but still required.

Parity Bit

A bit added to the byte as a check. Using odd parity, the bits must add up to an odd number, even parity must add up to an even number. No parity means that no extra bits have been added. (See Parity).

PBBS (Packet Bulletin Board System)

See BBS

PC (Personal Computer)

Usually refers to a computer that is identical in function to a product by IBM that was marketed as an IBM PC. PC can also mean any kind of computer that is used by an individual for general purposes (i.e. not a microwave oven control panel). It is sometimes hard to determine if a person who mentions PC is referring to a generic personal computer (i.e. Macintosh, Amiga, IBM PC, Atari, etc..) or a specific meaning one of these general computer processors. There are dozens of choices for packet radio software available for the PC-compatible. There are even several different custom hardware products that let you have a TNC built into the PC-compatible.

Ping (TCP/IP)

Command that allows the user to determine if another station is on the air, and tells the user how long it takes to send a message and get a response back.

PMS (Personal Mail System)
See Mail Box

Point-to-Point Link
This is a path between two sites. A point-to-point link is efficient, in terms of spectrum usage, because directional antennas may be used and signal gain maximized at each end of the link.

Poll Packet
In the AX.25 packet protocol, if a transmitted information packet is not acknowledged, the transmitting TNC will generate a poll packet to see if the destination TNC is still operating. If the poll packet is acknowledged then the transmitting TNC will once again attempt to send the information packet.

Prefix
The name of the volume(disk) or the letters before the number in callsigns.

Port
Pathway for transmission or reception of data. It represents a connector included on the back of a computer where you would plug in a cable to attach a peripheral to the computer. In packet terms, ports are where devices (i.e. TNC) are attached to allow communications out. Nodes and BBS system can support more than one port.

Protocol
A set of rules that govern the transmitting and receiving of data for the exchange of information between terminal and computer or between two computers. Protocols specify how a message is to be packaged: the sending address, the receiving address, error checks, etc. Many different protocols can be used, but all components within the same system must use the same protocol in order to communicate successfully with one another. AX.25 is a protocol which describes how small computers will talk to each other.

PROM
Programmable Read Only Memory. (See EPROM)

Protected Backbone
A protected backbone is a backbone where none of the known devices involved in the backbone will accept traffic from any unknown device. (See Backbone)

Quality
Net/Rom and TheNET software allows for a factor called quality. The quality factor is used to determine the path length for a network connect. Quality factors are determined as part of the node broadcast sequence. Each node sends a one way message out every half hour (setable) that tells its neighbors what nodes are listed in the nodes table. The neighbors have a quality factor that associates incoming node listings to new quality values.

RAM (Random Access Memory)

This is an IC in a computer that holds data only so long as power is applied. This is usually used only for storage during the execution of a program. TNCs use RAM for temporary storage, messages and parameters. Normally, TNC RAMs are powered all the time using a lithium battery in the TNC.

Real Time

When a signal is sent, a result is expected back within a short enough time. Keyboard to keyboard operation is real time. Keyboard to server is real time. Reading your mail from a remote BBS is real time. Sending a message via a packet BBS is not real time, because the sender doesn't know how long it will be before the friend's answer comes back.

Redundancy Path

In a perfect packet network, more than one path would exist between every two points. If one of the two paths is preferred due to load handling capability or number of node hops then that path would be called the primary path. The other path would be called the redundancy path.

Response Time

This is the time between sending data to a remote device before an expected response returns to the originating station.

RETRY

This is a process by which a packet that is not acknowledged is regenerated.

Reverse Forward

This is a feature of all modern BBSs and some PMSs where a connecting BBS may request if any mail is available to be taken by the connecting BBS. The prompt and answer exchange that takes place is in plain text and may be monitored if you are curious.

ROSE (RATS Open System Environment)

This is a networking software package created by Tom Moulton, W2W, in concert with the Radio Amateur Telecommunications Society in New Jersey, which implements a multiport, multistation packet radio network. The software consists of several parts. The most major part is in the form of an EPROM that resides in a TNC2 compatible TNC. Other parts include network management and system operation tools that run on a PC-compatible but which are not integral to the network's day-to-day operation. ROSE operation is done with the use of site callsigns and numeric designations that are traditionally in the form of a telephone area code and local exchange. The user treats the local ROSE 'switch' as a digipeater with the destination switch's numeric code as a second digipeater in the user connect sequence. The routing methodology used in ROSE is based on a fixed table stored in RAM in each switch and downloaded by the designated sysop.

Routing

Processes of sending data between network nodes. May be a direct path, use AX.25 digipeaters, NET/ROM nodes, or an IP station as a gateway.

Routing Loop

This is a condition where a packet is sent through a node or BBS more than once due to routing errors. In some instances, this only delays the time until the packet or message reachers it destination.

RS-232

See EIA-232

Serial Data Transmission

Communication where data is transmitted one bit at a time over one wire, as compared with parallel communication, where data is transmitted several bits at a time over several wires in parallel.

Serial port

An external connector on a computer that is used to connect a modem or other serial device. The typical serial port uses a DB-25 or DB-9 connector, based on the EIA-232 standard serial connection. Some computers may require additional hardware for a serial port to be available for use.

Server

A server is any stations that is operating as a service to other users than the owner. This may included BBSs, DX Packet Clusters, DOSgates, network nodes, etc.

SIG

Special Interest Group

Site Hardening

Term for ruggedizing a site by adding backup power, shielding, lightning protection, and other such fail safe devices. Although many devices don't help during 'acts of god.'

Site Manager

This is the person or persons that are responsible for node hardware and site access.

Site Sponsor

This is the person or persons who are financially involved in acquiring and maintaining node hardware.

Site Sysop

This is the person or persons who have software control over a node site.

SLIP (Serial Link Internet Protocol)

This protocol allows an IP station to connect via the serial port. Used with TCP/IP.

SMTP (Simple Mail Transfer Protocol)

This protocol allows the transfer of mail between TCP/IP stations. This is a non real time service which operates in a way very similar to normal packet BBSs.

Standards, De Facto

When a vendor's product is widely used, it becomes a de facto standard. Apple, IBM, Intel, Lotus, Hayes, Microsoft, Motorola and many other hardware and software vendors have set de facto standards.

Store-and-Forward System

Systems that store a message for transmission to its destination at a later time. Most network mail systems utilize store-and-forward techniques when direct communication is not available. Packet BBSs use this method.

SSID (Secondary Station IDentification)

A callsign is normally used as an address with the SSID of 0 (i.e. WD5IVD-0) In a case where a ham needs to have more than one address on the air at the same time the callsign may be used with additional SSIDs. There are 16 different possible SSIDs. WD5IVD-2, WD5IVD-3, WD5IVD-4 are examples of three different packet addresses, all using the same callsgin.

Stream

AX.25 allows many connections to be made from one station at the same time. Each connection is called a stream. The origin and address callsign pair for the connections must be different for each stream. That means if you have a connection to WD5IVD-1, you cannot connect to WD5IVD-1 on another stream a second time. This process is called multi-connect or streaming and is available on most modern TNCs. Look at the USERS command in your TNC's manual.

Synchronous Access

Accessing information in a real-time manner. The exchange of information requires that both or all parties must be present on the system at the same time to exchange information. Examples of synchronous access are on-line services (CIS, Prodigy), telephone calls, electronic conferences.

Sysop (SYStem OPerator)

An individual or individuals who run a system or bulletin board. The SYSOP may also act as mediator for system conferences.

TCP/IP (Transmission Control Protocol/Internet Protocol)

A set of communications protocols developed for the Defense Advanced Research Projects Agency (DARPA) to internetwork dissimilar systems. It runs on a large number of VAXs and UNIX-based computers. It is used by many corporations and almost all American universities and federal organizations. The File Transfer Protocol (FTP) and Simple Mail Transfer Protocol (SMTP) provide file transfer and e-mail capability. The TELNET protocol provides a terminal emulation capability that allows a user to interact with any other type of computer in the network. The TCP protocol controls the transfer of the data and the IP protocol provides the routing mechanism. Internet uses TCP/IP.

Telecommunications
Communication of all forms of information, including voice and video over telephone or computer networks.

Telecommunications Software
The program that allows your computer to transmit and receive messages.

Telecommuting
Working from home via computer; substitutes telecommunications and computers for the commute to work.

Telematics
The convergence of telecommunications and information processing.

Terminal
They are also referred to as dumb terminals. Sometimes a computer is used as a terminal. A terminal is usually a display screen and keyboard hooked to an EIA-232 port. When you type on the keyboard data is sent out of the TxData pin of the EIA-232 connector on the terminal. When RxData signals are detected on the EIA-232 connector the text is displayed on the screen.

TexNet
A packet-radio network developed by the Texas Packet Radio Society. TexNet consists of a combined hardware and firmware solution. TexNet is designed to be a dedicated, remote sited, multi-access, multi-resource network system. Users typically access network nodes using standard 1200 baud AFSK. This allows any user to have all the advantages of highspeed network operations without the requirement of purchasing additional station equipment beyond the normal TNC. The Texas TexNet network operates network links at 9600 baud.

Time To Live
In many networks, when a packet is sent from one network node to another network node, the packet contains several bytes of information which are useful for the intervening nodes. One of those bytes of information is time-to-live. Each time a node relays the packet one hop further the time-to-live is decremented. When it decrements to zero the message is discarded. Thus, if the number of hops that the packet has to go to reach it's specified destination is greater than the initial time-to-live, the packet won't get there. In addition, if the time-to-live on the return trip is not high enough, an acknowledgment will not be returned.

TheNET
This is a networking software package created by Hans Giese, DF2AU, in concert with NORD><LINK in Germany, which implements a multiport, multistation packet radio network, based on NetRom networking styles.

The major part of TheNet is in the form of an EPROM that resides in a TNC2 compatible TNC. Other parts include node configuration tools that run on a PC compatible but which are not required after initial system setup. TheNET operation is done with the use of site callsigns and mnemonic designations that are traditionally in the form of a city name. The user treats the local TheNET node as a remote command processor by first connecting to it, then away from it.

Throughput
This term specifies how many bytes sent by an origin station actually reached the destination station in a give period of time. Throughput is a much better number to describe network performance than baud rate. Baud rate only describes the number of bit transitions that may possibly leave a transmitter in a second. Throughput is a statistic that actually means something to an end user. Throughput is calculated either by observation or by taking the original baud rate, given in bytes per second, subtracting out all of the wasted time and overhead due to network protocols, the lost time due to congestion and collisions.

TNC (Terminal Node Controller)
This is the brains that connects the user's terminal (computer) to his radio so that he can communicate to other stations. The TNC's job is to take text typed on the terminal or computer and store it until the user hits a return. At that time, the text is sent to the destination station. Each line of text (ended with a carriage return) becomes a packet and is stored in the TNC until it can be sent to the destination station. The TNC also has commands that let the user set the callsign and other options. The TNC acts like a home phone modem, in that it converts digital character data to tones. The big difference between a TNC and a phone modem is that the TNC has the intelligence to direct your traffic to a specific destination and to receive connects using it's own microprocessor and internal software, while a phone dial a number to make its connection. By changing the internal software, TNCs may also be used for other purposes besides home station use. This includes running as part of a set of TNCs in a network node.

TTL (Transistor-Transistor Logic)
This is the name for logic circuits which operate using 0V for a zero and 5V for a one to do binary operations.

UART (Universal Asynchronous Receiver/Transmitter)
This is an IC which is used in a computer to construct a serial port.

Unproto
An unproto packet is a packet transmitted without expecting a response. Technically, it is called a UI frame, which means Unnumbered Information frame (i.e. packet). If a packet station were to send a beacon for all to see or were to call CQ, the station would use an Unproto packet.

User Channel

The frequency designated for users to access the network or service (i.e. BBS, DX Cluster). This frequency would be devoid of servers and other nodes aside from the one designated. A common error in congested areas is that services and users use the same channel, thus defeating the purpose of user-only channels.

Upload

To transfer information (files) from your computer 'up' to a host (i.e. BBS) computer.

WAN (Wide Area Network)

WANs are defined by the network creator. In most cases, WANs are collections of MANs or LANs that cover a large geographic area in relationship to the LAN or MAN.

Wireline Link

This is a connection between the modem headers of a pair or more of TNCs such that the TNCs communicate via their radio ports but without an intervening pair of radios. Because the modems are bypassed the TNCs may talk at a higher data rate than 1200. Links that involve commercial landline facilities between the TNCs or network nodes (thus the units are separated by a really long pair of wires) is called a wireline link.

Xon-Xoff (CtrlS/CtrlQ)

In communications, a simple asynchronous protocol that keeps the receiving device in synchronization with the sending device. When the buffer in the receiving device is full, it sends an x-off signal (transmit off) to the sending device, telling it to stop transmitting. When the receiving device is ready to accept more, it sends the sending device an x-on signal (transmit on) to start again. Xon-Xoff is used by computers and intelligent terminals to control the flow of data between the systems. Over packet networks and connections, other methods are used to achieve the same effect.

References:

Finke, C. R. (Ed.) (1992). TPRS Quarterly Report. Texas Packet Radio Society, Inc.

Horzepa, Stan. WA1LOU. (1989). Your Gateway to Packet Radio, 2nd Edition. ARRL, Newington, CT.

Jones, Greg. WD5IVD. (1991). Glossary of Technical Terms for TENET Master Training Session.

Snodgrass , Jack. kf5mg@kf5mg.#dfw.tx.usa.na, 44.28.0.14. (1992). Packet Radio General Information. Texas Packet Radio Society, PO Box 50238, Denton, TX, 76206-0238.

Lucas, Larry, Greg Jones, David Moore. (1992) An Educator's Alternative to Costly Telecommunications. Texas Center for Educational Technology, University of North Texas.

Torborg , Tadd. KA2DEW. (1992). NAPRA Notebook. Northwest Amateur Radio Association, Box 70405, Bellevue, WA, 98007.

Digital Groups (1/1/95)

Amateur Radio Research and Development Corp (AMRAD)
PO Box 6148
McLean, VA 22106-6148
Newsletter: AMRAD Newsletter

American Radio Relay League (ARRL)
225 Main St
Newington, CT 6111
Internet: INFO@ARRL.ORG
Newsletter: QEX / Gateway

Arizona Packet Radio Association
8402 E Angus Dr
Scottsdale, AZ 85251

Central Illinois Packet Radio User Society (CIPRUS)
c/o Larry Keeran K9ORP
RR 1 Box 99B
Downs, IL 61736-9717

Central Iowa Technical Society (CITS)
c/o Ralph Wallio W0RPK
1250 Hwy G24
Indianapolis, IA 50125

Chicago Amateur Packet Radio Association (CAPRA)
PO Box 8251
Rolling Meadows, IL 60008
Newsletter: The CAPRA Beacon

Cincinnati Amateur Packet Radio Experimenters Society (CAPRES)
c/o John Schroer IV KA8GRH
948 Halesworth Dr
Forest Park, OH 45240

Colorado Digital Electronics (CODE)
3631 Brentwood Terrace
Colorado Springs, CO 80910
Internet: info@code.org

Colorado Packet Association (COPA)
c/o John Radomski KT0H
2080 S Fairplay
Aurora, CO 80014

Connecticut Digital Radio Association (CONNECT)
c/o Tom Hogerty, KC1J
83 Harold Rd
Farmington, CT 6032

Eastern Packet Radio of Michigan (EPROM)
c/o Jay Nugent WB8TKL
3081 Braeburn Cir
Ann Arbor, MI 48104

EastNet BackBone Network (EBN)
c/o John Gubernard, K2LSX
81 Harcourt Ave
Bergenfield, NJ 07621-1916

Florida Amateur Digital Communications Association (FADCA)
812 Childers Loop
Brandon, FL 33511
Newsletter: FADCA>Beacon

Georgia Radio Amateur Packet Enthusiast Society (GRAPES)
PO Box 871
Alpharetta, GA 30239-0871
Newsletter: Grapevine

Hamilton and Area Packet Network (HAPN)
Box 4466 Station D
Hamilton, ON L8V 4S7
Canada

HEX 9 Group
PO Box 151
Orilla, ON L3V 6J3
Canada

Mid-Atlantic Packet Radio Club
c/o Tom Clark W3IWI
6388 Guilford Rd
Clarksville, MD 21029

Mississippi Amateur Radio Digital Association (MARDA)
c/o Patrick J Fagan WA5DYV
2412 E Birch Dr
Gulfport, MS 39503

Mt Ascutney Amateur Packet Radio Association
c/o Carl Breuning N1CB
54 Myrtle St
Newport, NH 03773

Mt Beacon Amateur Radio Club
PO Box 841
Wappingers Falls, NY 12590

New England Packet Radio Association (NEPRA)
PO Box 208
East Kingston, NH 03827
Newsletter: NEPRA PacketEar

North East Digital Association (NEDA)
PO Box 563
Manchester, NH 03105

Northern California Packet Association (NCPA)
6608B Alhambra Ave Suite 111
Martinez, CA 94553
Newsletter: NCPA Download

Northwest Amateur Packet Radio Association (NAPRA)
c/o John Gates N7BTI
750 Northstream Ln
Edmonds, WA 98020
Newsletter: Zero Retries

Ohio Packet Enthusiasts Club (OPAC)
c/o Bob Ball WB8WGA
830 Riva Ridge Blvd
Gahanna, OH 43230

Oregon Digital Network Coordination Council (ODNCC)
7860 SW 69th Av
Portland, OR 97223

Pacific Packet Radio Society (PPRS)
PO Box 51562
Palo Alto, CA 94303

Packet Radio Organization of Montana (PROM)
c/o Glenda Allen KE7TB
165 Conifer Rd
Libby, MT 59923

Packeteers of Long Island (POLI)
c/o Alex Mendelsohn AI2Q
92 Hathaway Av
Elmont, NY 11003
Newsletter: The POLI Parrot

Pennsylvania Packet Association (PaPA)
c/o Bryan Simanic WA3UFN
9 Wild Cherry Dr
DuBois, PA 15801

Radio Amateur Satellite Corp (AMSAT)
PO Box 27
Washington, DC 20044
Newsletter: AMSAT Journal

Radio Amateur Telecommunications Society (RATS)
c/o J Gordon Beattie Jr N2DSY
206 North Vivyen St
Bergenfield, NJ 07621

Rochester Packet Group
c/o Fred Cupp W2DUC
27 Crescent Rd
Fairport, NY 14450

San Diego Packet Group (SDPG)
c/o Mike Brock WB6HHV
10230 Mayer Cir
San Diego, CA 92126

San Diego Packet Radio Association (SANDPAC)
c/o Barry Gershenfeld
5085 Arroyo Lindo Av
San Diego, CA 92117
Newsletter: San Diego Packet Radio Association Newsletter

South Carolina Amateur Radio Digital Society (SCARDS)
PO Box 1281
Columbia, SC 29202
Newsletter: SCARDS Newsletter

Southern Amateur Packet Society (SAPS)
c/o Wayne Harrell WD4LYV
Rt 1 Box 368
Sycamore, GA 31790

Southern California Digital Communications Council (SCDCC)
PO Box 2744
Huntington Beach, CA 92647-2744
Newsletter: The I-Frame

Tennessee Network (TENNET)
c/o Jeffrey Austen K9JA
2051 Clearview Drive
Cookeville, TN 38506
Internet: jra1854@tntech.edu

Texas Packet Radio Society (TPRS)
PO Box 50238
Denton, TX 76206-0238
Internet: dwolf@tcet.unt.edu
Newsletter: The TPRS Quarterly Report

Tucson Amateur Packet Radio Corporation (TAPR)
8987-309 E. Tanque Verde Rd #337
Tucson, AZ 85749-9399
Internet: TAPR@TAPR.ORG
Newsletter: Packet Status Register

TwinsLAN Amateur Radio Club
c/o Kermit Kramer W0RFD
1121 Xerxes Av S
Minneapolis, MN 55405
Newsletter: The TwinsLAN Beacon

Utah Packet Radio Association (UPRA)
c/o Bart Van Allen KA7ZFD
11883 S Kinney Cir
Riverton, UT 84065

Vancouver Amateur Digital Communications Group (VADCG)
9531 Odlin Rd
Richmond, BC V6X 1E1
Canada
Newsletter: The Packet

Wake Digital Communications Group (WDCG)
c/o Randy Ray WA5SZL
9401 Taurus Ct
Raleigh, NC 27612

Western Michigan Packet Radio Association (WMPRA)
PO Box 4612
Muskegon, MI 49444

Wisconsin Amateur Packet Radio Association (WAPRA)
PO Box 1215
Fond Du Lac, WI 54935
Newsletter: Badger State Smoke Signals

Index

E

F

G

H

I

K

L

M

TAPR needs you!

Zero in on the international organization that started the packet revolution in the early '80s. TAPR continues today to work on new technology, standards, and publications. Combine this work with the Packet Status Register, an authoritative source since 1982 for up-to-date user and technical information about packet radio, and you can't go wrong when you join TAPR!

Don't be left out! TAPR is the place to get connected on a national and regional basis. Find out what others are doing and what *you* might do next in digital communications!

Goals:

- Support R&D in amateur digital communications
- Disseminate information on digital communications
- Provide affordable and useful kits
- Pursue and help advance the amateur art of communications through publications, meetings, and standards

History:

If you have used a packet radio TNC, then you are already a part of TAPR history.

The TNC (Terminal Node Controller) project grew from a discussion in October of 1981. This was the genesis of TAPR. The project progressed from these first prototype units to the TNC-1 and then eventually to the TNC-2, which is the basis for most amateur packet operations worldwide.

TAPR was founded in 1982 as a national organization with interests in the areas of packet and digital communications. Today, TAPR continues as a membership supported non-profit [501(c)(3)] amateur research and development organization. TAPR currently has more than 2500 members, worldwide. TAPR continues to develop kits for the amateur community and is working actively on publications and communications standards

Membership:

Membership in TAPR is $15 for one year. In Canada and Mexico membership is $18 and outside North America, $25.

Join the amateur digital revolution. Join TAPR!

TAPR Product Overview

January 1995

TAPR has been active in the development of amateur radio kits since its origination. What follows is a description of current TAPR kits. A current price list can be attained by contacting the TAPR office or via the Internet.

State Machine DCD Upgrade

For TNCs with modems such as:

KPC-1, KPC-4, KAM	PacComm Tiny-2/Micro-2
DRSI PC*PA Type 1 & 2	KPC-2400
AIWA APX-25, APX-25M	AEA PK-87, 88, 232
TASCO TNC-20, 20H	PacComm TNC-220
Heath HK-232	Heath Pocket Packet

The upgrade adaptor for these TNCs adds an EPROM-based State Machine to derive DCD based on lockup of a digital phase-lock loop. It is a small PC board assembly that mounts easily inside the cabinet of most any TNC (**NOT** the Heath Pocket Packet/TASCO TNC-μ21). This upgrade will **dramatically** improve DCD operation, allowing you to run your radio unsquelched. This alone reduces other stations' TXDelay requirements, improving channel throughput. DCD Clock Option: For TNCs with "software HDLC," an optional clock generator may be included to provide the necessary clock signal (KPC-1, KPC-2, HF channel of KAM).

XR-2211 DCD Upgrade

For TNCs using the XR-2211 demodulator such as:

TAPR Beta Board	GLB PK-1, TNC-2A
TNC-1, TNC-2	PacComm TNC-200
Heath HD-4040	MFJ 1270, 1274
AEA PKT-1, PK-80	DRSI HF*MODEM

The PC board is tiny and shaped to fit into a TNC-1 or TNC-2. After construction (an hour or so), you simply unplug the XR-2211 chip from its socket, insert it into the socket on the upgrade board, then plug the upgrade board into the IC socket vacated by the XR-2211 chip on the TNC.

If you are into HF operation, provisions are made to connect a "Threshold" control onto the demodulator. The result will be fast-attack, slow decay DCD with a hang time to compensate for temporary fades due to multipath. When all stations sharing a channel have proper DCD action, data flows more efficiently.

AN-93 HF RTTY, AMTOR, PCTOR Modem

The AN-93 kit provides any PC user with the capability for operating RTTY, AMTOR, and PCTOR with this simple modem-only design. AN-93 is the equivalent of a BayCom, BayPac, or PMP setup, but for HF digital operations. This very simple kit should be ideal for many who have wanted to play on HF, but didn't want to pay the money for an expensive multi-mode controller. With the AN-93, only three components are required for HF digital communications: a PC-compatible computer, the AN-93 modem, and software that performs the encoding and decoding. The AN-93 comes with a tuning indicator to allow visual tuning and the unit also provides audio output for oscilloscope display. The AN-93 requires 9volts @ 150mA. Power Supply is not included in the kit. Interfacing to the radio is through a DB-9. TAPR provides a simple method tuning the modem. The kit will be shipped with the A/D convertor providing full-memory ARQ capability for PCTOR. The kit provides both FSK and AFSK outputs for use with more common amateur HF radios.

PK-232 Modem Disconnect

This kit is a simple PC board assembly you can build in an evening and still have time left to operate your radio! Once completed, it simply plugs into an IC socket inside the PK232 and provides you with a "Standard" TAPR modem disconnect header. This header provides a convenient place to connect the DCD State Machine upgrade kit, the TAPR 9600bps modem, PSK modem, or other items that need this connection. The result is a no-holes enhancement of your PK-232 (or Heath HK-232).

TAPR Software Library

In addition to supplying various kits and firmware, TAPR maintains a library of packet radio-related computer software. Disks are currently available in 3-1/2 in. for $3.00 each, including mailing (slightly more for foreign orders). Some stock remains for 5-1/4 in. MS-DOS format for $2.00 each, and is noted on the TAPR Order Form. TAPR attempts to provide the latest versions of all software.

TAPR 9600bps Modem

NOTE: This kit requires some prior hands-on experience with radio modifications.

The TAPR 9600bps modem is a full-duplex baseband modem compatible with K9NG and G3RUH standards. The TAPR 9600bps modem, like other 9600bps modems, requires an interface to your radio other than via the microphone and speaker jacks. The modem fits internally into the case of many popular TNCs and supports a 20-pin modem disconnect header which allows for a second external modem (such as the TAPR PSK modem) to optionally be connected.

The transmit path includes an independent watchdog timer. The modem selection logic correctly routes the PTT command from the TNC to prevent, for example, keying your 1200 baud radio when running 9600bps. The receive filter includes a compensation adjustment to tailor the modem to the radio receiver and AC-coupling to accommodate frequency drift on the channel. An improved state machine is used for clock recovery. The DCD circuit works on both signal quality and the presence of a signal.

9600bps Modem Clock Option provides the necessary radio port clock for 9600bps operation. When you enable the external 9600bps modem on a TNC 1, TNC 2, or clone, your TNC will automatically be set up for the correct radio port speed. When you switch back to the TNC's internal modem, the TNC will automatically select its normal radio port speed again. Thus, you can switch between 1200bps and 9600bps operation with a single switch. The clock option is necessary if you wish to use your modem as part of a stand-alone full-duplex 9600bps repeater, in conjunction with the bit regenerator option.

9600bps Modem Bit Regenerator Option

This option is intended only for use as part of a full-duplex 9600bps repeater. It is not suitable for use at a home station. The bit regenerator keys the PTT line of the modem whenever it receives an incoming signal (DCD is true). It stores the first 8 bits in a buffer, then starts sending the bits out again, resynchronizing them with its own crystal-controller clock to remove jitter and noise from the received signal. Full-duplex bit-regenerating repeaters for 9600bps are starting to be used in more and more locations. It truly makes 9600bps operation fast and friendly.

DSP-93

TAPR AMSAT®

The TAPR/AMSAT DSP-93 is designed to provide radio amateurs the wonderful capabilities of Digital Signal Processing in a stand-alone low-cost design. Not just limited to one mode, the DSP-93 can support data, audio, and video modes with the proper software. The DSP-93 is supplied as a complete kit (including box and power supply). *For those not wanting to build a kit, there are several preassembled DSP units on the market today.*

DSP-93 has been designed in a modular fashion with two four-layer boards utilizing an interconnecting bus structure. The DSP Engine, bottom board, contains the TMS320C25 DSP, 32K by 16 bits of program and data memory - upgradable to 64K, the clock circuitry (40Mhz) and some programmable array logic for system I/O. The Radio/Computer Interface Board, top board, contains two eight-pin female mini-DIN connectors for radio interfacing. The Texas Instruments TLC32044 Analog I/O chip is used, which samples and updates at a rate of up to 45K operations per second and includes aliasing filters. This board also communicates to your computer at speeds up to 19.2K baud using a serial connection.

DSP-93, as of 1/95, has the following software available: 1200 baud AFSK, 300 baud AFSK, 300 baud Packet HF AFSK, 1200 PSK, 9600 FSK terrestrial, 9600 FSK full-duplex for satellite operations, audio filters, HF modem (for RTTY, AMTOR, PCTOR), APT, SSTV, satellite 400 bps telemetry, and much more is being developed. User interface software packages for Macintosh and Windows is available, each provides the interface for a digital scope, a spectrum analyzer and function generator. A low-cost shareware assembler is available for code development. Source code is made available for most of the code provided with the unit.

The DSP-93 sells for $430 as a complete kit (1/95), including enclosure and power supply. If you live in a country that does not use 110v 60Hz AC, the price is $420, and the unit will ship without the 9v AC wall transformer. Please inform the TAPR office when you order if you do not require the wall transformer. All other parts will be in the kit, ready for complete assembly. The kit is done twice yearly, contact the TAPR office about the next production.

NOSIntro $ 23.00

An Introduction to the KA9Q Network Operating System. Great for the beginning NOS or TCP/IP user. Ian Wade, G3NRW

TAPR's Packet Radio Book $12.00

Articles covering beginning & intermedtiate packet usage. Intro Stuff, Beginners' Frames, Networking, BBSs, and more!

ARRL Computer Networking Conf. & Digital Communications Conf. Proceedings

1st - 4th	(1981, '83, '84, '85)		$ 18.00
5th	(1986)	Orlando, FL	$ 10.00
6th	(1987)	Redondo Beach, CA	$ 10.00
7th	(1988)	Columbia, MD	$ 12.00
8th	(1989)	Colorado Springs, CO	$ 12.00
9th	(1990)	London, ONT Canada	$ 18.00
10th	(1991)	San Jose, CA	$ 10.00
11th	(1992)	Teaneck, NJ	$ 10.00
Entire Set	(1981 - 1992)		$ 80.00

Future TAPR Titles Include:

- BBS Sysop Operational Guide
- Land Mobile Radio Modifications for 9600 baud
- NOS TCP/IP Primer

TAPR Members receive 10% discount on all kits and publications.

TAPR • 8987-309 E Tanque Verde Rd #337 Tucson, Az • 85749-9399 • (817) 383-0000
TAPR is a Non-Profit R & D Corporation

○ **Send me more information on TAPR and its activities.**

○ **I want to JOIN TAPR and get two years membership for $25, a savings of $5.00**

Name: ____________________ Call: ____________

Address: ____________________ Apt: ____________

____

City: ____________ State: ________ Zip: ________

Home Phone: ( ) ____________ Work Phone: ( ) ____________

Credit Card# (Visa/MC) ____________ Expires: ________

• Number of years: ________ • Is this a renewal? (Yes/No): ________

Tucson Amateur Packet Radio
8987-309 E. Tanque Verde Rd #337
Tucson, Arizona • 85749-9399
Office: (817) 383-0000 • Fax: (817) 566-2544
Non-Profit Research and Development Corporation